T0222351

Editorial Policy

§ 1. Lecture Notes aim to report new developments - quickly, informally, and at a high level. The texts should be reasonably self-contained and rounded off. Thus they may, and often will, present not only results of the author but also related work by other people. Furthermore, the manuscripts should provide sufficient motivation, examples and applications. This clearly distinguishes Lecture Notes manuscripts from journal articles which normally are very concise. Articles intended for a journal but too long to be accepted by most journals, usually do not have this "lecture notes" character. For similar reasons it is unusual for Ph. D. theses to be accepted for the Lecture Notes series.

§ 2. Manuscripts or plans for Lecture Notes volumes should be submitted (preferably in duplicate) either to one of the series editors or to Springer- Verlag, Heidelberg . These proposals are then refereed. A final decision concerning publication can only be made on the basis of the complete manuscript, but a preliminary decision can often be based on partial information: a fairly detailed outline describing the planned contents of each chapter, and an indication of the estimated length, a bibliography, and one or two sample chapters - or a first draft of the manuscript. The editors will try to make the preliminary decision as definite as they can on the basis of the available information.

§ 3. Final manuscripts should preferably be in English. They should contain at least 100 pages of scientific text and should include
- a table of contents;
- an informative introduction, perhaps with some historical remarks: it should be accessible to a reader not particularly familiar with the topic treated;
- a subject index: as a rule this is genuinely helpful for the reader.

Further remarks and relevant addresses at the back of this book.

Lecture Notes in Mathematics 1687

Editors:
A. Dold, Heidelberg
F. Takens, Groningen
B. Teissier, Paris

Springer
*Berlin
Heidelberg
New York
Barcelona
Budapest
Hong Kong
London
Milan
Paris
Santa Clara
Singapore
Tokyo*

Folkmar Bornemann

Homogenization in Time of Singularly Perturbed Mechanical Systems

 Springer

Author

Folkmar Bornemann
Lehrstuhl für Numerische Mathematik
und Wissenschaftliches Rechnen
Technische Universität München
D-80290 München, Germany
e-mail: bornemann@na-net.ornl.gov

Cataloging-in-Publication Data applied for

Die Deutsche Bibliothek - CIP-Einheitsaufnahme

Bornemann, Folkmar A.:
Homogenization in time of singularly perturbed mechanical systems /
Folkmar Bornemann. - Berlin ; Heidelberg ; New York ; Barcelona ;
Budapest ; Hong Kong ; London ; Milan ; Paris ; Santa Clara ;
Singapore ; Tokyo : Springer, 1998
  (Lecture notes in mathematics ; 1687)
  ISBN 3-540-64447-4

Mathematics Subject Classification (1991):
34Cxx, 34Exx, 35Qxx, 70Kxx, 81Q15

ISSN 0075-8434
ISBN 3-540-64447-4 Springer-Verlag Berlin Heidelberg New York

Typesetting: Camera-ready TeX output by the author
SPIN: 10649830   46/3143-543210 - Printed on acid-free paper

in memoriam Alfred Neumann (1902–1982)

Preface

Although the title might suggest it differently, this monograph is about a certain method for establishing singular limits rather than about a clear-cut class of singularly perturbed problems. Using this particular method I will address in a unified way such diverse topics as the micro-scale justification of the Lagrange-d'Alembert principle and the limit behavior of strong constraining potentials in classical mechanics on the one hand, and the adiabatic theorem of quantum mechanics on the other hand. I am confident that all these topics are *cases* of a larger class of singularly perturbed mechanical systems that show up rapid micro-scale fluctuations, allowing for the application of the method to be presented. Reflecting this, I have tried to apply the method to each case as directly as possible and refrained from studying an abstract super-class of problems which would leave the cases as mere examples. I believe that this "variations-on-a-theme-style" of my presentation is more likely to make the method a working tool in the area than a "transformation-to-an-archetype-style" would be.

This monograph grew out of the attempt to understand the high frequency vibrations in classical molecular dynamics modeling. These nonlinear vibrations are the major obstruction for an efficient and reliable numerical long term simulation. In the fall of 1994, I came up with the idea of studying the singular limit of these vibrations by means of the *method of weak convergence* which enjoys growing popularity in the study of singularly perturbed nonlinear partial differential equations. Straightforward energy arguments led me to a qualitative understanding of the structural aspects of the limit system. However, additional ideas appeared to be necessary for the *explicit* construction of the limit dynamics. I finally found these ideas in the physical concepts of *virial theorems* and *adiabatic invariants*. For the *single frequency* case weak limit analogues and proofs were discovered soon, and the method was presented in the spring of 1996. In contrast, the *multiple frequency* case left two problems open: first, how to obtain a kind of component-wise virial theorem, and second, how to get rid of a certain perturbation term obstructing the multidimensional adiabaticity argument. More than a year of struggle later, I discovered quite elegant solutions to both of these problems: the virial theorem generalizes to a matrix-commutativity relation[†] that, after simultaneous diagonalization,

---

[†]To be found as Lemma II.7 on p. 38 below.

implies the desired component-wise result. And the mentioned perturbation term vanishes as a consequence of a resonance condition by a strikingly short argument.[‡]

In the meantime, I got involved in the study of mixed quantum-classical models in quantum chemistry. During the fall of 1996, my friend and colleague Christof Schütte suggested to me to discuss the singular limit of a finite dimensional analogue of these models by transforming it to the kind of classical mechanical systems that had already been studied by my method. However, the kinetic energy of the transformed system turned out to be of a more general type as considered before, making it necessary to generalize my results to mechanical systems on Riemannian manifolds. The Riemannian metric caused additional perturbation terms which, surprisingly enough, vanished like magic because of the already discovered generalization of the virial theorem.

Encouraged by this success I worked on a direct, untransformed version of the method of proof for these mixed quantum-classical models. The motivation was to deal with the infinite dimensional case involving partial differential equations. And, indeed, using appropriate concepts from physics and the right tools from functional analysis, I have not only been able to address this case but also to give a new proof for the adiabatic theorem of quantum mechanics.

All this endeavor shaped the methodological point of view which I will pursue in this monograph: *case studies*. I hope that the method presented here, i.e., the blend of weak convergence techniques, virial theorems, and adiabatic invariants, will find many interesting new applications and will help to establish, clarify, and unify results about singularly perturbed problems involving different time scales.

New York, April 1997   *Folkmar A. Bornemann*

[‡]To be found as Lemma II.11 on p. 42 below.

Acknowledgments[§]

It gives me pleasure to thank all those individuals who made this work possible by providing support, encouragement, advice, criticism, teaching, and friendship. Out of the many I would like to mention a few to whom I am especially thankful:

Olof Widlund, for inviting me to a one-year stay at the Courant Institute of Mathematical Sciences in New York. Relieving me from any obstructing obligations, he made possible an enjoyable and scientifically profitable visit at this lively place with its own, uniquely stimulating research atmosphere.

Peter Deuflhard, for his continuing support and promotion of many years. His foresighted interest in a contribution of applied mathematics to bio-technologies by the efficient numerical simulation of bio-molecules initiated the beginnings of this work.

Christof Schütte, for fruitful and mutually stimulating discussions, for his encouragement and criticism. My work on the subject of this monograph started, in fact, as a joint enterprise with him, witnessed by our joint publications on the single frequency case and on the mixed quantum-classical modeling issue.

Luc Tartar and François Murat, for their beautifully instructive and stimulating series of lectures presented at the "DMV-Seminar on Composite Materials and Homogenization, Oberwolfach, Germany, March 12–18, 1995," which I had the unique opportunity to attend.[¶]

Carles Simó, for posing various important problems and instructive questions after reading an earlier manuscript of mine. His suggestion to include gyroscopic force terms stimulated the study of §II.5. Also, by asking for a more detailed comparison with the averaging techniques available in perturbation theory of integrable Hamiltonian systems, he initiated the case study of Appendix C.

And those, whose courses at the Freie Universität Berlin, some ten years ago, taught me several of the analytical skills which were essential for accomplishing this work: Volker Enß (real analysis), Michael Loss (differential geometry), and Dirk Werner (functional analysis).

[§]My research on the subject of this monograph was supported in part by the U.S. Department of Energy under contract DE-FG02-92ER25127.

[¶]This seminar was organized by the German mathematical society (Deutsche Mathematiker-Vereinigung).

Contents

I

Introduction

Many problems of the applied sciences involve scales in time, or space, which are orders of magnitude different in size. The smallest scales, also called *micro-scales*, are caused, e.g., by dynamical effects or by the materials involved, whereas the largest scales, also called the *macro-scales*, involve the scales of observation or measurement. The micro-scales are frequently not measurable, or at least of no particular interest. Additionally, their presence poses severe problems for numerical simulations using todays or even future computing facilities. Thus, for first a deeper understanding of the underlying model, and second, for developing efficient and reliable numerical simulation methods, there is a strong need for macro-scale models which approximate the originally given model *without* involving the micro-scales.

Looking at many macro-scale models used in the natural sciences, one realizes that they were obtained *phenomenologically*, i.e., by analyzing the measurements, not by *deriving* them from a micro-scale model. For justifying well-known macro-scale models, or establishing even new ones, techniques for a mathematical model-derivation from the micro- to the macro-scale have become increasingly important and scientifically instructive in recent years.

A fundamental mathematical technique used in such a kind of model-derivation is the identification of a scale-parameter $\epsilon \ll 1$ and the subsequent study of the limit $\epsilon \to 0$. Frequently, this limit changes, at least *formally*, the very nature of the mathematical model: Either some terms become formally ambiguous, or even non-sense, or, e.g., the order or type of a differential equation is changed. In these cases, one calls the model *singularly perturbed*, and ϵ the parameter of the singular perturbation. The analysis of the *singular limit* $\epsilon \to 0$ has to be either *asymptotic* or, in a sense, *oblivious* to the micro-scale aspects of the solution.

In some models, singular perturbations cause rapid, micro-scale fluctuations in the solution. Therefore, an *asymptotic description* usually involves an explicit ansatz for the oscillatory part of the solution. This often requires much ingenuity and a lot of insight into the problem. Famous examples are provided by the *perturbation theory of integrable Hamiltonian systems*, cf. [6], the *WKB method* for semiclassical limits in quantum theory, cf. [68], the more recent technique of *nonlinear geometric optics*, cf. [67][75], and the method of *multiple-scale-asymptotics*, cf. [13][53].

On the other hand, macro-scale measurements can be viewed as a kind of averaging procedure, being oblivious of the rapid fluctuations on micro-scales. Mimicking this, one considers the limit of certain averages of the solutions or, equivalently, their *weak limits*. This *method of weak convergence* has become increasingly popular in the study of nonlinear partial differential equations, since one has powerful tools from functional analysis at hand which allow to establish qualitative, or at least structural, information about the limit system, cf. [28]. Even more, proving error bounds for a formal asymptotic (multiple-scale) analysis can be a hard problem which is sometimes attacked by the method of weak convergence.

Example 1. Both approaches to a singular limit problem can neatly be compared being applied to the so-called *homogenization* problem of elliptic partial differential equations. There, one studies the limit $\epsilon \to 0$ of the diffusion problems

$$-\operatorname{div} A(x/\epsilon) \operatorname{grad} u_\epsilon(x) = f(x), \quad x \in \Omega \subset \mathbb{R}^d, \qquad u_\epsilon|_{\partial\Omega} = 0,$$

where the $[0,1]^d$-periodic diffusion matrix $A \in L^\infty(\mathbb{R}^d, \mathbb{R}^{d\times d})$ describes a micro-scale periodic structure. For $f \in H^{-1}(\Omega)$, we obtain

$$u_\epsilon \rightharpoonup u_0 \quad \text{in} \quad H^1(\Omega),$$

and one might ask whether there is an *effective* diffusion matrix $A_{\text{eff}} \in L^\infty(\Omega, \mathbb{R}^{d\times d})$ such that

$$-\operatorname{div} A_{\text{eff}}(x) \operatorname{grad} u_0(x) = f(x), \quad x \in \Omega \subset \mathbb{R}^d, \qquad u_0|_{\partial\Omega} = 0.$$

It turns out that such an effective diffusion matrix exists and is constant indeed. However, this matrix is in general *not* simply given by some average value of $A(\cdot)$. For this problem, a multiple-scale-analysis was set up by BENSOUSSAN, LIONS, and PAPANICOLAOU [13]. The weak convergence method was pioneered by MURAT and TARTAR [70][47]. In the latter reference one can also find a proof of an error bound for the multiple-scale-expansion by means of the weak convergence method, cf. [47, §1.4].

The particularity of this example is provided by a *nonlinear coupling*, in fact a quadratic one, of the micro- and macro-scales, which leads to nontrivial problems and counter-intuitive results in general. This effect of nonlinearities can be understood in a rather direct fashion by the method of weak convergence: nonlinear functionals are *not* weakly sequentially continuous in general.[1] Describing the deviation from weak continuity, unexpected terms appear in the weak limit. In analogy to the elliptic example above, we will use the notion "homogenization," instead of the notion "averaging," if the derivation of a macro-scale model involves such unexpected, counter-intuitive, or nontrivial terms.

---

[1]For instance, in Example 1 above, the flux $A(x/\epsilon) \operatorname{grad} u_\epsilon(x)$ is the product of two weakly converging sequences. Thus, although we have $A(x/\epsilon) \rightharpoonup A_{\text{aver}}$ in $L^\infty(\Omega, \mathbb{R}^{d\times d})$ and $\operatorname{grad} u_\epsilon \rightharpoonup \operatorname{grad} u_0$ in $L^2(\Omega, \mathbb{R}^d)$, there is *not* $A(x/\epsilon) \operatorname{grad} u_\epsilon(x) \rightharpoonup A_{\text{aver}} \operatorname{grad} u_0$ in $L^2(\Omega, \mathbb{R}^d)$. This way, one understands why $A_{\text{aver}} \neq A_{\text{eff}}$.

Purpose of this Monograph. It is our purpose in this monograph to present a particular method for the explicit homogenization of certain singularly perturbed mechanical systems. Caused by the singular perturbation and properties of the model, the solutions of these systems will show up rapid micro-scale fluctuations. Our method will be based on energy principles and weak convergence techniques. Since nonlinear functionals are *not* weakly sequentially continuous, as mentioned above, we have to study *simultaneously* the weak limits of all those nonlinear quantities of the rapidly oscillating components which are of importance for the underlying problem.[2] Using the physically motivated concepts of *virial theorems*, *adiabatic invariants*, and *resonances*, we will be able to establish sufficiently many relations between all these weak limits, allowing to calculate them explicitly.

Our approach will be to *exemplify* by a *case study* rather than aiming at the largest possible generality. This way, we can show most clearly how concepts and notions from the physical background of the underlying mathematical problem enter and help to determine relations between weak limit quantities.

Example 2. We present a simplified version of the problem that will be discussed in Chapter II. It is given by the singularly perturbed Newtonian equations

$$\epsilon^2\, \ddot{x}_\epsilon + \operatorname{grad} U(x_\epsilon) = 0,$$

describing a mechanical system on the configuration space $\mathbb{R}^m$. If the potential $U \geq 0$ has a critical manifold

$$N = \{x \in \mathbb{R}^m : U(x) = 0\} = \{x \in \mathbb{R}^m : \operatorname{grad} U(x) = 0\},$$

we call it *constraining* to N. For fixed initial values $x_\epsilon(0) = x_* \in N$ and $\dot{x}_\epsilon(0) = v_* \in \mathbb{R}^m$, we obtain

$$x_\epsilon \xrightarrow{*} x_0 \quad \text{in} \quad W^{1,\infty}([0,T], \mathbb{R}^m)$$

where the limit average motion x_0 takes values in the manifold N. Therefore, naïve intuition would expect that the limit x_0 is dynamically described by the free, geodesic motion on N,

$$\ddot{x}_0 \perp T_{x_0} N.$$

---

[2]In this restriction to just the smallest required class of nonlinearities our approach differs from utilizing and studying *Young measures* [28, §1.E.3][97], or *H-measures* (*microlocal defect measures*) [37][95][96][97] and its scale-dependent variants like *semiclassical measures* [36] or *Wigner measures* [65]. These advanced tools encode the weak limits of *all possible* quantities obtained from nonlinear substitutions or quadratic pseudodifferential operations, respectively. For the relation of our approach to semiclassical measures see Appendix D.

However, the rapidly oscillating velocities, being only weakly convergent in general, cause additional, unexpected force terms which yield a limit dynamics of the form

$$\ddot{x}_0 + \operatorname{grad} U_{\mathrm{hom}}(x_0) \perp T_{x_0} N.$$

We call this limit system the *homogenization* of the given singularly perturbed system. The nonlinearity responsible for the appearance of the homogenized potential U_{hom} is the *quadratic* approximation of U near to N. We will construct U_{hom} *explicitly* for a large class of potentials U.

Outline of Contents. For pedagogical reasons, in §2 of this introductory Chapter I we will introduce our method of proof for a very simple, illustrative special case of Example 2 above. This will help to clarify the basic structure of the argument. Precedingly, we recall in §1 all those prerequisites about weak convergence which will be needed in the first three chapters of this monograph.

The first major case for the application of our method is subject of Chapter II. There, we study natural mechanical systems on Riemannian manifolds, singularly perturbed by a strong constraining potential. We state and prove a homogenization result which considerably extends what is known from the only two references concerned with this problem: i.e., work by TAKENS [94] from 1979, and by KELLER and RUBINSTEIN [52] from 1991. We show the necessity of resonance conditions on the one hand, but verify that genericity or transversality assumptions are sufficient on the other hand.[3] As a particular subcase we discuss the micro-scale justification of the Lagrange-d'Alembert principle by utilizing strong constraining potentials. Giving unified proofs for the results known about this justification, we additionally show the necessity of the conditions which prior to our work were only known to be sufficient. The chapter concludes with an explicit example that there is a strange sensitivity of the homogenization problem on the initial values, if the strong potential does not satisfy an important regularity assumption. For this effect was discovered by TAKENS [94], we call it *Takens chaos*.

Chapter III continues with a potpourri of applications. We start by discussing the problem of *guiding center motion* in plasma physics and the elimination of fast vibrations in *molecular dynamics*. The closing application consists of a simplified, finite-dimensional version of a model in quantum chemistry which describes the coupling of a quantum mechanical system with a classical one. We show that this model can be transformed to the case of Chapter II.

---

[3]The appearance of resonance-conditions for proving adiabaticity in multi-frequency systems is a well-known fact in the perturbation theory of integrable Hamiltonian systems, cf. [6, §5.4.2]. However, the resonance conditions employed there are far too restrictive for our purposes, see also Appendix C.

The corresponding infinite-dimensional coupling model, in its original, untransformed guise leads to the second major case of this monograph, subject of Chapter IV. There, we relate the singular limit of the coupling model to the adiabatic theorem of quantum mechanics. By translating our method of proof to the appropriate concepts of quantum theory, we give a new proof for this theorem and, finally, discuss the singular limit of the coupling model. This way, we find ourselves in the curious situation having addressed adiabaticity in classical and quantum mechanics with essential the same method of proof, thus adding new meaning to the correspondence principle of Ehrenfest in the pre-Schrödinger, "old" quantum theory of the early twenties.

As the proofs in Chapter IV have to deal with operator-valued function spaces and parameter-dependent unbounded operators they require the application of more sophisticated tools from functional analysis than needed in the preceding chapters. These tools are not easily found and referenced in the literature, so that we have to provide some proofs for the reasons of logical completeness. However, for not obstructing the inherent simplicity of the basic argument in Chapter IV, we placed all the more general functional analytic material in Appendix B.

§1. The Basic Principles of Weak Convergence

Here, we present all those facts from real and functional analysis about weak* convergence that will be applied in the first three chapters.[4] We group these facts into *five basic principles*, all well-known or trivial. The experienced reader who is already familiar with these concepts might wish to skip this section.

We consider sequences $\{x_\epsilon\}$ of functions, indexed by a sequence $\{\epsilon\}$ of real numbers which converge to zero, $\epsilon \to 0$. Later on, ϵ will be the singular perturbation parameter. All functions x will be defined on some bounded Lipschitz domain $\Omega \subset \mathbb{R}^d$, the expression ∂x will stand for any partial derivative $\partial_j x$, $j = 1, \ldots, d$.

We recall the fact [84, Theorem 6.16] that the function space $L^\infty(\Omega)$ is the *dual* of the function space $L^1(\Omega)$,

$$L^\infty(\Omega) = (L^1(\Omega))^*.$$

Thus, the functional analytic concept of weak*-convergence specifies as follows.

Definition 1. A sequence $\{x_\epsilon\}$ of $L^\infty(\Omega)$ *converges weakly** to the limit $x_0 \in L^\infty(\Omega)$, notated as $x_\epsilon \overset{*}{\rightharpoonup} x_0$, if and only if

$$\int_\Omega x_\epsilon(t)\phi(t)\, dt \to \int_\Omega x_0(t)\phi(t)\, dt \qquad \text{as } \epsilon \to 0,$$

---

[4]Additional material needed for Chapter IV can be found in Appendix B.

for all testfunctions $\phi \in L^1(\Omega)$.[5]

There is an alternative way of describing this kind of convergence which clarifies the connection to *averaging* or *filtering* concepts.

Lemma 1. *A sequence $\{x_\epsilon\}$ of $L^\infty(\Omega)$ converges weakly* to a limit $x_0 \in L^\infty(\Omega)$, if and only if the following two properties hold:*

(i) *the sequence is bounded in $L^\infty(\Omega)$,*

(ii) *for every open rectangle $I \subset \Omega$ the corresponding integral mean value converges,*

$$\frac{1}{|I|} \int_I x_\epsilon(t)\, dt \;\to\; \frac{1}{|I|} \int_I x_0(t)\, dt.$$

Proof. The linear span of the characteristic functions χ_I forms a dense subspace of $L^1(\Omega)$, cf. [98, §11, Theorem 4] and [98, §20, Theorem 4]. This proves the "if"-part of the the assertion, since by property (i) the sequence $\{x_\epsilon\}$ represents a sequence of equi-continuous linear forms on $L^1(\Omega)$.

The "only if" part follows from the uniform boundedness principle of functional analysis [83, Theorem 2.5].   □

The weak* convergence in $L^\infty(\Omega)$ helps to ignore rapid fluctuations in a convenient way. As an example, we recall the well-known Riemann-Lebesgue lemma [84, Sec. 5.14]: For $x_\epsilon(t) = \cos(t/\epsilon)$ one gets

$$x_\epsilon \overset{*}{\to} 0 \quad \text{in} \quad L^\infty[0,1]. \tag{I.1}$$

The first principle, simple but extremely useful, relates the uniform convergence of functions to the weak* convergence of their derivatives.

Principle 1. *Let $\{x_\epsilon\}$ be a sequence in $C^1(\overline{\Omega})$ such that $x_\epsilon \to 0$ in $C(\overline{\Omega})$. Then, if and only if the sequence $\{\partial x_\epsilon\}$ is bounded in $L^\infty(\Omega)$, there holds*

$$\partial x_\epsilon \overset{*}{\to} 0 \quad \text{in} \quad L^\infty(\Omega).$$

Proof. By the uniform boundedness principle [83, Theorem 2.5], a weakly* convergent sequence must be bounded, proving the necessity of the boundedness condition. On the other hand, because $C_c^\infty(\Omega)$ is a dense subspace of $L^1(\Omega)$, [63, Lemma 2.19], one can test for the weak* convergence of a bounded sequence in $L^\infty(\Omega)$ with functions from $C_c^\infty(\Omega)$. Now, since by Hölder's inequality uniform convergence in $C(\overline{\Omega})$ implies weak* convergence in $L^\infty(\Omega)$, we obtain by partial integration

$$\partial x_\epsilon(\phi) = -x_\epsilon(\partial\phi) \;\to\; 0$$

---

[5]We will write, for short, $x(\phi) = \int_\Omega x(t)\phi(t)\, dt.$

for all testfunctions $\phi \in C_c^\infty(\Omega)$, proving the sufficiency of the boundedness condition.   □

For instance, this principle provides quite a simple proof of the Riemann-Lebesgue lemma, Eq. (I.1) above: Since $\epsilon \sin(t/\epsilon) \to 0$, uniformly in $C[0,1]$, the *uniformly bounded* sequence of its derivatives, $\cos(t/\epsilon)$, converges weakly* to zero, $\cos(t/\epsilon) \overset{*}{\to} 0$.

A delicate point about weak* convergence is that nonlinear functionals are *not* weakly* sequentially continuous. Given a *nonlinear* continuous function $f : \mathbb{R} \to \mathbb{R}$, there is, in general,

$$x_\epsilon \overset{*}{\to} x_0 \quad \not\Rightarrow \quad f(x_\epsilon) \overset{*}{\to} f(x_0).$$

A famous example is provided by the quadratic function $f(x) = x^2$ and $x_\epsilon = \cos(t/\epsilon)$. There, by a well-known generalization of the Riemann-Lebesgue lemma, [23, Lemma I.1.2], we have that in $L^\infty[0,1]$

$$x_\epsilon \overset{*}{\to} 0, \qquad \text{but} \qquad x_\epsilon^2 \overset{*}{\to} \frac{1}{2\pi} \int_0^{2\pi} \cos^2 \tau \, d\tau = \frac{1}{2} \neq 0. \qquad \text{(I.2)}$$

This *lack* of weak* sequential continuity is responsible for a lot of unexpected results later on. It makes the study of singular perturbation problems with rapid fluctuations a difficult but fascinating task.

However, if the weak* convergence enters an nonlinear expression only linearly, passing to the limit is possible. A particular important instance is provided by our second principle. This principle will be used most often in our work, and always *without referring* to it explicitly. Therefore, if we claim a passage to the weak* limit for a product, the reader is cautioned to check *carefully* whether *at most one* factor is weakly* converging.

Principle 2. *Let there be the convergences $x_\epsilon \overset{*}{\to} x_0$, weakly* in $L^\infty(\Omega)$, and $y_\epsilon \to y_0$, uniformly in $C(\overline{\Omega})$. Then, we obtain*

$$x_\epsilon \cdot y_\epsilon \overset{*}{\to} x_0 \cdot y_0 \quad \text{in} \quad L^\infty(\Omega).$$

Proof. Take $\phi \in L^1(\Omega)$. Then, Lebesgue's theorem of dominated convergence shows that $y_\epsilon \phi \to y_0 \phi$ strongly in $L^1(\Omega)$. Therefore, we get

$$(x_\epsilon \cdot y_\epsilon)(\phi) = x_\epsilon(y_\epsilon \phi) \to x_0(y_0 \phi) = (x_0 \cdot y_0)(\phi),$$

which proves the asserted weak*-convergence.   □

The next two principles establish a particularly convenient property of the weak*-topology: the *Heine-Borel property*, i.e., closed bounded sets are compact. Since the underlying topology is essentially induced by a metric, weak* convergence is almost as easy to handle as convergence in $\mathbb{R}$: bounded sequences have convergent subsequences.

Principle 3. (Alaoglu Theorem). *Let $\{x_\epsilon\}$ be a bounded sequence in the space $L^\infty(\Omega)$. Then, there is a subsequence $\{\epsilon'\}$ and a function $x_0 \in L^\infty(\Omega)$, such that*

$$x_{\epsilon'} \overset{*}{\rightharpoonup} x_0 \quad \text{in} \quad L^\infty(\Omega).$$

Proof. The Alaoglu theorem [83, Theorem 3.15] of functional analysis states that a closed ball in $L^\infty(\Omega)$ is compact with respect to the weak*-topology. Since the predual space $L^1(\Omega)$ is separable, cf. [63, Lemma 2.17], the weak*-topology is *metrizable* on closed balls [83, Theorem 3.16]. Hence, bounded sequences have weak*-convergent subsequences.[6]   □

The next compactness principle is the classical Arzelà-Ascoli theorem, slightly extended to include some information on the derivative.[7]

Principle 4. (An Extended Arzelà-Ascoli Theorem). *Let $\{x_\epsilon\}$ be a bounded sequence in the space $C^{0,1}(\overline{\Omega})$ of uniformly Lipschitz continuous functions. Then, there is a subsequence $\{\epsilon'\}$ and a function $x_0 \in C^{0,1}(\overline{\Omega})$, such that*

$$x_{\epsilon'} \to x_0 \quad \text{in} \quad C(\overline{\Omega}), \qquad \partial x_{\epsilon'} \overset{*}{\rightharpoonup} \partial x_0 \quad \text{in} \quad L^\infty(\Omega).$$

The partial derivatives ∂x_ϵ and ∂x_0 are classically defined almost everywhere.

Proof. The uniform bound on the Lipschitz constants of x_ϵ implies the *equi-continuity* of the sequence. Thus, the classical Arzelà-Ascoli theorem [84, Theorem 11.28] shows that there is a subsequence ϵ' and a continuous function $x_0 \in C(\overline{\Omega})$, such that $x_{\epsilon'} \to x_0$ in $C(\overline{\Omega})$.

By Rademacher's theorem [29, Theorem 3.1.2], a Lipschitz function x is differentiable almost everywhere. The thus obtained derivatives ∂x belong to L^∞ and equal the weak derivatives of x in the sense of distributions. This way, one gets (e.g., [29, Theorem 6.2.1] or [38, p. 154]) an isometric isomorphism

$$C^{0,1}(\overline{\Omega}) \cong W^{1,\infty}(\Omega).$$

Hence, the sequences $\{\partial x_\epsilon\}$ are bounded in $L^\infty(\Omega)$. By Principle 3, we can choose the subsequence $\{\epsilon'\}$ in such a way that $\partial x_{\epsilon'} \overset{*}{\rightharpoonup} \eta_0$ in $L^\infty(\Omega)$. Now, the very same argument as in the proof of Principle 1 shows that $\eta_0 = \partial x_0$. In particular, the above isomorphism implies that $x_0 \in C^{0,1}(\overline{\Omega})$ which finishes the proof.   □

By these compactness results, bounded sequences of functions turn out to be a *mixture* of sequences which converge in appropriate topologies. The

---

[6]RUDIN [84, Theorem 11.29] gives a direct proof of this *sequential* version of the Alaoglu theorem—i.e., for spaces with a *separable predual*—which requires little more than the Arzelà-Ascoli theorem [84, Theorem 11.28]. This sequential version is often called the *Banach-Alaoglu theorem.*

[7]It can be viewed as the Alaoglu theorem for the space $C^{0,1}(\overline{\Omega}) \cong W^{1,\infty}(\Omega)$ together with the *compact* Sobolev embedding $W^{1,\infty}(\Omega) \hookrightarrow C(\overline{\Omega})$.

•

fifth and final principle states a simple but very useful criterion for deciding
about the convergence of the given sequence itself.

Principle 5. ("Uniqueness Implies Convergence"). *Let $\{x_\epsilon\}$ be a
sequence in a sequentially compact Hausdorff space $\mathscr{X}$. If every convergent
subsequence of $\{x_\epsilon\}$ converges to one and the same element $x_0 \in \mathscr{X}$, then
the sequence converges itself,*

$$x_\epsilon \to x_0.$$

Proof. Suppose on the contrary that x_ϵ does not converge to x_0. Then,
there is a neighborhood $U(x_0)$ of x_0 and a subsequence $\{\epsilon'\}$ such that

$$x_{\epsilon'} \notin U(x_0) \qquad \text{for all } \epsilon'. \tag{*}$$

Since $\mathscr{X}$ is sequentially compact, there is a subsequence $\{\epsilon''\}$ of $\{\epsilon'\}$ such
that $\{x_{\epsilon''}\}$ converges. By assumption $x_{\epsilon''} \to x_0$, a contradiction to (*). □

As has been explicitly stated in the proof of Principle 3, the weak*-
topology of L^∞ is *metrizable* on bounded sets. Thus, the convergence
Principle 5 is applicable to the bounded sequences of Principles 3 and 4.

§2.   An Illustration of the Method

Here, we study a simple illustrative model problem which will serve as the
skeleton of the arguments in Chapters II and IV. After having studied this
model problem, the reader will more easily enjoy the inherent simplicity of
our method which tends to be hidden by several technical difficulties later
on. For instance, the technical difficulties in Chapter II are due to the differential geometrical setting, in Chapter IV due to the infinite dimensional
spaces and unbounded operators.

We have arranged this illustrative model problem in a way that the four
basic steps of the argument will be clearly visible:

1. weak* compactness based on an energy principle,

2. a weak virial theorem,

3. the adiabatic invariance of the normal actions, and

4. the identification of the limit mechanical system.

Surely, there would be short-cuts for this particularly simple problem
which, however, are worth being sacrificed for a presentation of as many
essential features and notions of the later proofs as possible.

§2.1. The Model Problem

We consider the following singularly perturbed system of Newtonian equations of motion,

$$\epsilon^2 \ddot{x}_\epsilon + \operatorname{grad} U(x_\epsilon) = 0, \tag{I.3}$$

describing a mechanical system with Euclidean configuration space $M = \mathbb{R}^m$. Splitting the coordinates according to $x = (y, z) \in \mathbb{R}^n \times \mathbb{R}^r = \mathbb{R}^m$, we specify the potential U by the quadratic expression

$$U(x) = \tfrac{1}{2}\langle H(y)z, z \rangle, \qquad H(y) = \operatorname{diag}(\omega_1^2(y), \ldots, \omega_r^2(y)).$$

Here, $\langle \cdot, \cdot \rangle$ denotes the Euclidean inner product on $\mathbb{R}^r$; correspondingly, $|\cdot|$ will denote Euclidean norms. We assume that the *smooth* functions ω_λ are uniformly positive,[8] i.e., there is a constant $\omega_* > 0$ such that

$$\omega_\lambda(y) \geq \omega_*, \qquad y \in \mathbb{R}^n, \quad \lambda = 1, \ldots, r.$$

The nonnegative potential $U \geq 0$ is called to be *constraining* to the *critical submanifold*

$$N = \{x \in M : U(x) = 0\} = \mathbb{R}^n \times \{0\} \subset M.$$

Thus, for obvious reasons, we call y the *tangential*, and z the *normal* component of x. A component-wise writing of the equations of motion, Eq. (I.3), yields

$$
\begin{aligned}
\text{(i)} \quad \ddot{y}_\epsilon^j &= -\tfrac{1}{2}\epsilon^{-2}\langle \partial_j H(y_\epsilon)z_\epsilon, z_\epsilon \rangle, \qquad j = 1, \ldots, n, \\
\text{(ii)} \quad \ddot{z}_\epsilon &= -\epsilon^{-2} H(y_\epsilon)z_\epsilon.
\end{aligned}
\tag{I.4}
$$

We consider initial values which are independent of ϵ,

$$y_\epsilon(0) = y_*, \quad \dot{y}_\epsilon(0) = w_*; \quad z_\epsilon(0) = 0, \quad \dot{z}_\epsilon(0) = u_*. \tag{I.5}$$

Notice, that the particular choice $z_\epsilon(0) = 0$ is the only one that results in an ϵ-independent bound for the conserved energy E_ϵ of the system; in fact, making the energy even independent of ϵ,

$$E_\epsilon = \tfrac{1}{2}|\dot{y}_\epsilon|^2 + \tfrac{1}{2}|\dot{z}_\epsilon|^2 + \epsilon^{-2}U(y_\epsilon, z_\epsilon) = \tfrac{1}{2}|w_*|^2 + \tfrac{1}{2}|u_*|^2 = E_*. \tag{I.6}$$

§2.2. Step 1: Equi-Boundedness (Energy Principle)

In this step, energy and compactness arguments allow us to extract appropriately converging quantities. Conservation of energy and H being positive definite immediately imply that the velocities are uniformly bounded,

$$\dot{y}_\epsilon = O(1), \quad \dot{z}_\epsilon = O(1),$$

---

[8]This assumption is crucial, cf. Footnote 13 on p. 18 as well as Eqs. (II.6) and (II.10).

as functions in $C[0,T]$, given a certain final time $0 < T < \infty$. Therefore, after integration, we also obtain the positional bounds

$$y_\epsilon = O(1), \quad z_\epsilon = O(1).$$

The estimate

$$\tfrac{1}{2}\epsilon^{-2}\omega_*^2|z_\epsilon|^2 \le \tfrac{1}{2}\epsilon^{-2}\langle H(y_\epsilon)z_\epsilon, z_\epsilon\rangle = \epsilon^{-2}U(y_\epsilon, z_\epsilon) \le E_*$$

implies the uniform bound

$$z_\epsilon = O(\epsilon).$$

Inserting this into the first set (i) of the equations of motion (I.4) reveals the acceleration bound

$$\ddot{y}_\epsilon = O(1).$$

Now, an application of the extended Arzelà-Ascoli theorem, Principle 4, and the Alaoglu theorem, Principle 3, yields—after the extraction of a subsequence which we denote by $\epsilon \to 0$ again—the convergences

$$y_\epsilon \to y_0 \quad \text{in} \quad C^1([0,T], \mathbb{R}^n), \qquad \ddot{y}_\epsilon \overset{*}{\rightharpoonup} \ddot{y}_0 \quad \text{in} \quad L^\infty([0,T], \mathbb{R}^n),$$

$$\dot{z}_\epsilon \overset{*}{\rightharpoonup} 0 \quad \text{in} \quad L^\infty([0,T], \mathbb{R}^r), \qquad \epsilon^{-1}z_\epsilon \overset{*}{\rightharpoonup} \eta_0 \quad \text{in} \quad L^\infty([0,T], \mathbb{R}^r).$$

Multiplying the second set (ii) of the equations of motion (I.4) by ϵ and taking weak* limits gives, by recalling Principle 1,

$$0 = H(y_0)\eta_0, \quad \text{i.e.,} \qquad \eta_0 = 0.$$

Because quadratic functionals are not weakly* sequentially continuous in general, we cannot expect that the uniformly bounded matrices

$$\Sigma_\epsilon = \epsilon^{-2}z_\epsilon \otimes z_\epsilon, \qquad \Pi_\epsilon = \dot{z}_\epsilon \otimes \dot{z}_\epsilon,$$

likewise converge weakly* to the zero matrix. Instead, after a further application of the Alaoglu theorem, Principle 3, and an extraction of subsequences, we obtain *some* limits Σ_0 and Π_0,

$$\Sigma_\epsilon \overset{*}{\rightharpoonup} \Sigma_0 \quad \text{in} \quad L^\infty([0,T], \mathbb{R}^{r\times r}), \qquad \Pi_\epsilon \overset{*}{\rightharpoonup} \Pi_0 \quad \text{in} \quad L^\infty([0,T], \mathbb{R}^{r\times r}).$$

These limit quantities Σ_0 and Π_0 will play a crucial role in the description of the limit dynamics of y_0. As a first hint, we rewrite the first set (i) of the equations of motion (I.4) as

$$\ddot{y}_\epsilon^j = -\tfrac{1}{2}\operatorname{tr}(\partial_j H(y_\epsilon) \cdot \Sigma_\epsilon).$$

By taking weak* limits on both sides, we obtain what we call the *abstract limit equation* for y_0,

$$\ddot{y}_0^j = -\tfrac{1}{2}\operatorname{tr}(\partial_j H(y_0) \cdot \Sigma_0). \tag{I.7}$$

The next two steps establish sufficiently many relations between the limits y_0, Σ_0, and Π_0, so that the force term of the abstract limit equation can be expressed as a function of the limit y_0 *alone*.

§2.3. Step 2: The Weak Virial Theorem

First, we establish a relation between Π_0 and Σ_0. The matrix

$$\Xi_\epsilon = \dot{z}_\epsilon \otimes z_\epsilon = O(\epsilon)$$

converges uniformly to the zero matrix. Therefore, taking weak* limits of its time derivative,

$$\dot{\Xi}_\epsilon = \dot{z}_\epsilon \otimes \dot{z}_\epsilon + \ddot{z}_\epsilon \otimes z_\epsilon = \Pi_\epsilon - H(y_\epsilon)\Sigma_\epsilon$$

yields, by Principle 1, the equation

$$0 = \Pi_0 - H(y_0)\Sigma_0. \tag{I.8}$$

In particular, the *diagonal* entries of Σ_0 and Π_0 are related by

$$\Pi_0^{\lambda\lambda} = \omega_\lambda^2(y_0)\Sigma_0^{\lambda\lambda}. \tag{I.9}$$

This result is essentially about the energy distribution in the normal, oscillating component z_ϵ. The second set (ii) of the equations of motion (I.4) implies that each normal component z^λ ($\lambda = 1, \ldots, r$) satisfies the equation of a *fast harmonic oscillation* whose frequency is *slowly perturbed*,

$$\ddot{z}_\epsilon^\lambda + \epsilon^{-2}\omega_\lambda^2(y_\epsilon)z_\epsilon^\lambda = 0. \tag{I.10}$$

We define the *kinetic energy* $T_{\epsilon\lambda}^\perp$ and the *potential energy* $U_{\epsilon\lambda}^\perp$ of the normal λ-component by

$$T_{\epsilon\lambda}^\perp = \tfrac{1}{2}|\dot{z}_\epsilon^\lambda|^2 = \tfrac{1}{2}\Pi_\epsilon^{\lambda\lambda}, \qquad U_{\epsilon\lambda}^\perp = \tfrac{1}{2}\epsilon^{-2}\omega_\lambda^2(y_\epsilon)|z_\epsilon^\lambda|^2 = \tfrac{1}{2}\omega_\lambda^2(y_\epsilon)\Sigma_\epsilon^{\lambda\lambda},$$

the *total energy* is given by $E_{\epsilon\lambda}^\perp = T_{\epsilon\lambda}^\perp + U_{\epsilon\lambda}^\perp$. The diagonal limit relation (I.9) implies that, in each component, the weak* limits of the kinetic and potential energy are *equal*,

$$T_{\epsilon\lambda}^\perp \overset{*}{\rightharpoonup} T_{0\lambda}^\perp = \tfrac{1}{2}\omega_\lambda^2(y_0)\sigma_\lambda, \qquad U_{\epsilon\lambda}^\perp \overset{*}{\rightharpoonup} U_{0\lambda}^\perp = \tfrac{1}{2}\omega_\lambda^2(y_0)\sigma_\lambda,$$

abbreviating $\Sigma_0^{\lambda\lambda} = \sigma_\lambda$. The limit of the total energy is

$$E_{\epsilon\lambda}^\perp \overset{*}{\rightharpoonup} E_{0\lambda}^\perp = \omega_\lambda^2(y_0)\sigma_\lambda. \tag{I.11}$$

The thus obtained *equi-partitioning* of energy into the kinetic and the potential part bears similarities with the virial theorem of classical mechanics. For this reason, we call the result (I.8) the *weak virial theorem*.

§2.4. Step 3: Adiabatic Invariance of the Normal Actions

Now, we establish a relation between Σ_0 and y_0. The normal *actions* are given by the energy-frequency-ratios

$$\theta_\epsilon^\lambda = \frac{E_{\epsilon\lambda}^\perp}{\omega_\lambda(y_\epsilon)}, \qquad \lambda = 1, \ldots, r.$$

We will show their *adiabatic invariance*, i.e., the uniform convergence

$$\theta_\epsilon^\lambda \;\to\; \theta_0^\lambda = \text{const}.$$

This will be accomplished by calculating the weak* limit of the time derivative $\dot E_{0\lambda}^\perp$ in a twofold way. On the one hand, by using (I.10), the time derivative of the energy $E_{\epsilon\lambda}^\perp$ is

$$\dot E_{\epsilon\lambda}^\perp = \dot z_\epsilon^\lambda \underbrace{\left(\ddot z_\epsilon^\lambda + \epsilon^{-2}\omega_\lambda^2(y_\epsilon)z_\epsilon^\lambda\right)}_{=0} + \tfrac{1}{2}\epsilon^{-2}|z_\epsilon^\lambda|^2\frac{d}{dt}\omega_\lambda^2(y_\epsilon).$$

Thus, the time derivatives are *bounded* functions and a further application of the extended Arzelà-Ascoli theorem, Principle 4, shows that the normal energies—and therefore the actions also—are in fact *uniformly* converging. Now, the weak limit of the time derivative is given by

$$\dot E_{\epsilon\lambda}^\perp \;\overset{*}{\rightharpoonup}\; \dot E_{0\lambda}^\perp = \tfrac{1}{2}\sigma_\lambda\frac{d}{dt}\omega_\lambda^2(y_0).$$

On the other hand, by a the direct differentiation of the limit (I.11) we obtain

$$\dot E_{0\lambda}^\perp = \dot\sigma_\lambda\,\omega_\lambda^2(y_0) + \sigma_\lambda\frac{d}{dt}\omega_\lambda^2(y_0).$$

A comparison of the two expressions obtained for $\dot E_{0\lambda}^\perp$ implies the equation of logarithmic differentials

$$\frac{\dot\sigma_\lambda}{\sigma_\lambda} = -\tfrac{1}{2}\frac{d\omega_\lambda^2(y_0)/dt}{\omega_\lambda^2(y_0)} = -\frac{d\omega_\lambda(y_0)/dt}{\omega_\lambda(y_0)}.$$

Solving this explicitly shows that there are *constants* θ_*^λ ($\lambda = 1,\dots,r$) such that

$$\sigma_\lambda = \frac{\theta_*^\lambda}{\omega_\lambda(y_0)}, \quad \text{i.e.,} \quad \theta_0^\lambda = \frac{E_{0\lambda}^\perp}{\omega_\lambda(y_0)} = \theta_*^\lambda.$$

The *values* of these constants can be calculated at the initial time $t = 0$,

$$\theta_*^\lambda = \theta_0^\lambda(0) = \lim_{\epsilon\to 0}\theta_\epsilon^\lambda(0) = \frac{|u_*^\lambda|^2}{2\,\omega_\lambda(y_*)}.$$

§2.5. Step 4: Identification of the Limit Mechanical System

Finally, we reconsider the limit force field on the right hand side of the abstract limit equation (I.7). By the results of the preceding section we obtain

$$\tfrac{1}{2}\operatorname{tr}(\partial_j H(y_0)\cdot\Sigma_0) = \tfrac{1}{2}\sum_{\lambda=1}^{r}\sigma_\lambda\,\partial_j\omega_\lambda^2(y_0) = \sum_{\lambda=1}^{r}\theta_*^\lambda\,\partial_j\omega_\lambda(y_0) = \partial_j U_{\text{hom}}(y_0).$$

Here, we define the *homogenized potential* U_{hom} by

$$U_{\text{hom}} = \sum_{\lambda=1}^{r} \theta_*^{\lambda} \, \omega_{\lambda}.$$

Notice that U_{hom}—and therefore y_0—does not depend on the chosen sub-sequences. By Principle 5, this allows us to discard the extraction of sub-sequences altogether. Summarizing, we have just proven the following theorem.[9]

Theorem 1. *Let y_{hom} be the solution of the second order differential equation*

$$\ddot{y}_{\text{hom}}^{j} = -\partial_j U_{\text{hom}}(y_{\text{hom}}), \qquad j = 1, \ldots, n,$$

with initial values $y_{\text{hom}}(0) = y_*$, $\dot{y}_{\text{hom}}(0) = w_*$. Then, for every finite time interval $[0, T]$, we obtain the strong convergence

$$y_{\epsilon} \to y_{\text{hom}} \quad in \quad C^1([0, T], \mathbb{R}^n)$$

and the weak* convergences $\epsilon^{-1} z_{\epsilon} \stackrel{*}{\rightharpoonup} 0$ and $\dot{z}_{\epsilon} \stackrel{*}{\rightharpoonup} 0$ in $L^{\infty}([0, T], \mathbb{R}^r)$.

§2.6.   Comments on the Notions Introduced and the Result

Let us comment on the notions "weak virial theorem" and "adiabatic invariance" as well as on one particularity of the result, Theorem 1.

Weak Virial Theorem.   There is a much deeper relation of what we called the weak virial theorem, Eq. (I.8), to the virial theorem of classical mechanics than just equi-partitioning of kinetic and potential energies. We follow the textbook of ABRAHAM and MARSDEN [1, p. 242ff.] for a recollection of the classical virial theorem. Given a vector field X on the configuration space M, the associated *momentum function* $P(X)$ is defined by

$$P(X) : TM \to \mathbb{R}, \qquad P(X)(v) = \langle X, v \rangle.$$

The *virial function* is defined by the Poisson bracket

$$G_{\epsilon}(X) = \{P(X), E_{\epsilon}\}.$$

The virial theorem [1, Theorem 3.7.30] states that the time average of the virial function along a trajectory is zero,

$$\frac{1}{T} \int_0^T G_{\epsilon}(X)(v_{\epsilon}) \, dt \to 0 \qquad \text{as } T \to \infty.$$

---

[9]For reasons of comparison, a further proof of this theorem, utilizing *asymptotic* techniques, is subject of Appendix C.

Here, $v_\epsilon = (\dot{y}_\epsilon, \dot{z}_\epsilon)$ denotes the velocity field of the trajectory under consideration. Now, we consider the following specific normal vector fields and momentum functions associated with:

$$X_{\lambda\mu} = z^\mu \frac{\partial}{\partial z^\lambda}, \qquad P(X_{\lambda\mu})(v_\epsilon) = z^\mu_\epsilon \dot{z}^\lambda_\epsilon, \qquad \lambda, \mu = 1, \ldots, r.$$

This yields the virial function

$$
\begin{aligned}
G_\epsilon(X_{\lambda\mu})(v_\epsilon) &= \sum_\nu \frac{\partial P}{\partial z^\nu} \frac{\partial E_\epsilon}{\partial \dot{z}^\nu} - \frac{\partial E_\epsilon}{\partial z^\nu} \frac{\partial P}{\partial \dot{z}^\nu} = \dot{z}^\lambda_\epsilon \frac{\partial E_\epsilon}{\partial \dot{z}^\mu} - z^\mu_\epsilon \frac{\partial E_\epsilon}{\partial z^\lambda} \\
&= \dot{z}^\lambda_\epsilon \dot{z}^\mu_\epsilon - \epsilon^{-2}(H(y_\epsilon)z_\epsilon)^\lambda \cdot z^\mu_\epsilon = (\Pi_\epsilon - H(y_\epsilon)\Sigma_\epsilon)^{\lambda\mu}.
\end{aligned}
$$

Thus, the assertion (I.8) can be rewritten in the form

$$G_\epsilon(X_{\lambda\mu})(v_\epsilon) \overset{*}{\rightharpoonup} 0 \qquad \text{as } \epsilon \to 0.$$

In fact, this far reaching analogy with the virial theorem of classical mechanics motivated the name *weak virial theorem*.

Adiabatic Invariance. The time interval $[0, T]$ under consideration is of the order $O(\epsilon^{-1}\tau_\epsilon)$, where τ_ϵ denotes a typical "period" of a small oscillation in the normal direction. Thus, the usage of the notion "adiabatic invariant" is in accordance with the usual definition given in textbooks on classical mechanics, as for instance in Arnold, Kozlov, and Neishtadt [6, Chap. 5.4]. In fact, the perturbation theory of integrable Hamiltonian systems is directly applicable for the single-frequency case $r = 1$. To this end, we rescale time and positions and introduce corresponding momenta

$$\tau = \epsilon^{-1}t, \qquad q = \epsilon^{-1}y, \qquad \eta = \epsilon^{-1}z, \qquad p = \dot{y}, \qquad \pi = \dot{z}.$$

Denoting derivatives with respect to the new time τ by a prime, Eq. (I.4) is just the canonical system

$$q'_\epsilon = \frac{\partial E_\epsilon}{\partial p_\epsilon}, \qquad p'_\epsilon = -\frac{\partial E_\epsilon}{\partial q_\epsilon}, \qquad \eta'_\epsilon = \frac{\partial E_\epsilon}{\partial \pi_\epsilon}, \qquad \pi'_\epsilon = -\frac{\partial E_\epsilon}{\partial \eta_\epsilon},$$

belonging to the energy E_ϵ as defined in Eq. (I.6) which, by suppressing the index '1,' transforms to

$$E_\epsilon = \tfrac{1}{2}|p_\epsilon|^2 + \tfrac{1}{2}\pi^2_\epsilon + \tfrac{1}{2}\omega(\epsilon q_\epsilon)\eta^2_\epsilon = \tfrac{1}{2}|p_\epsilon|^2 + \theta_\epsilon \omega(\epsilon q_\epsilon).$$

Now, a result of the perturbation theory of integrable Hamiltonian systems, [6, Chap. 5.4, Theorem 24 and Example 20], shows that for times $\tau = O(\epsilon^{-1})$ there is the asymptotics

$$\theta_\epsilon = \theta_* + O(\epsilon), \qquad \epsilon q_\epsilon = \epsilon q_0 + O(\epsilon), \qquad p_\epsilon = p_0 + O(\epsilon),$$

where the constant θ_* is defined by the initial values and q_0 and p_0 by the canonical equations of motion belonging to the limit energy function

$$E_0 = \tfrac{1}{2}|p_0|^2 + \theta_* \, \omega(\epsilon q_0).$$

These equations of motion are just the homogenized system of Theorem 1. Transforming back we thus obtain, for times $t = O(1)$, the error estimates[10]

$$\theta_\epsilon = \theta_* + O(\epsilon), \qquad y_\epsilon = y_{\text{hom}} + O(\epsilon), \qquad \dot{y}_\epsilon = \dot{y}_{\text{hom}} + O(\epsilon).$$

The Result. Later on, the reader will notice that the analogues of Theorem 1 in the more complicated situations of Chapter II and Chapter IV require certain *resonance conditions* to be imposed on the normal frequencies ω_λ. This is due to the following fact: general nonlinear potentials U, position dependent eigenspaces of H, and general manifolds M and constraints N introduce perturbations to the simple model of this introduction. Thus, instead of the harmonic oscillator equation (I.10) we will just get something like

$$\ddot{z}_\epsilon^\lambda + \epsilon^{-2}\omega_\lambda^2(y_\epsilon)z_\epsilon^\lambda = O(1).$$

While the equi-partitioning of the kinetic and potential energy still holds true then, the adiabatic invariance of the action might suffer from resonances. We illustrate this claim by the simple scalar equation

$$\ddot{z}_\epsilon + \epsilon^{-2}\omega^2 z_\epsilon = \cos(\epsilon^{-1}\omega t),$$

with a constant frequency $\omega > 0$. For the initial values $z(0) = \dot{z}(0) = 0$, we get the solution

$$z_\epsilon(t) = \tfrac{1}{2}\epsilon\omega^{-1}t\sin(\epsilon^{-1}\omega t).$$

By Eq. (I.2), the kinetic and potential normal energy are equal in the limit,

$$T_\epsilon^\perp(t) = \tfrac{1}{2}|\dot{z}_\epsilon(t)|^2 = \tfrac{1}{8}t^2\cos^2(\epsilon^{-1}\omega t) + O(\epsilon) \xrightarrow{*} T_0^\perp(t) = \tfrac{1}{16}t^2,$$

$$U_\epsilon^\perp(t) = \tfrac{1}{2}\epsilon^{-2}\omega^2|z_\epsilon(t)|^2 = \tfrac{1}{8}t^2\sin^2(\epsilon^{-1}\omega t) \xrightarrow{*} U_0^\perp(t) = \tfrac{1}{16}t^2.$$

On the other hand, the limit normal energy,

$$E_\epsilon^\perp(t) = \tfrac{1}{2}|\dot{z}_\epsilon(t)|^2 + \tfrac{1}{2}\epsilon^{-2}\omega^2|z_\epsilon(t)|^2 = \tfrac{1}{8}t^2 + O(\epsilon) \rightarrow E_0^\perp(t) = \tfrac{1}{8}t^2,$$

is *not* of the form $\theta_*\omega$, θ_* being a constant.

---

[10] Even better estimates can be found in Appendix C.

II

Homogenization of Natural Mechanical Systems with a Strong Constraining Potential

A *natural mechanical system* [6, p. 10] consists of a smooth Riemannian configuration manifold M with metric $\langle \cdot, \cdot \rangle$ and a smooth potential function $W : M \to \mathbb{R}$. The dynamics is described by the Lagrangian

$$\mathscr{L}(x, \dot{x}) = \tfrac{1}{2}\langle \dot{x}, \dot{x} \rangle - W(x), \qquad \dot{x} \in T_x M.$$

We will consider a family of singularly perturbed potentials of the form

$$W_\epsilon(x) = V(x) + \epsilon^{-2} U(x),$$

where the "strong" potential U is constraining to a smooth *critical submanifold* $N \subset M$. If we choose initial values with uniformly bounded energy, the solutions x_ϵ of the equations of motion are oscillating on a time-scale of order $O(\epsilon)$ within a distance of order $O(\epsilon)$ to the submanifold N. The sequence of solutions converges uniformly to some function x_0 of time taking values in N.

We will study the problem of a *dynamical description* for this limit x_0, i.e., whether there is a mechanical system with configuration space N such that x_0 is a solution of the corresponding equations of motion. We call this problem the *homogenization problem* for the given mechanical system.

For a large class of constraining potentials this homogenization problem admits a surprisingly elegant and explicit solution which we will state in §1. In §2 we will present a proof of the homogenization result based on the method of weak convergence. Compared to the existing literature,[11] this method of proof allows to weaken the imposed resonance conditions considerably.

The so-called problem of *realization of holonomic constraints* provides a special case of the homogenization problem. Here, one studies the question, whether the limit x_0 is just the solution of the equations of motion for the natural mechanical system with configuration space N and potential V. By the Lagrange-d'Alembert principle, the limit $\epsilon \to 0$ would then "realize" the holonomic (positional) constraint $x \in N$. In §3 we will establish necessary and sufficient conditions for this to happen, first, on the initial values for

---

[11] Short reviews of the existing literature can be found in §§1.10 and 3.3.

general constraining potentials U, and second, on the constraining potential U for general initial values.

If the constraining potential does not belong to the class introduced in §1, the limit dynamics can be of a completely different nature. In §4 we will present an explicit example for which the limit dynamics depends extremely sensitive on how the limit initial values are obtained. We will argue that in this case there is no solution of the homogenization problem which is comparably elegant to the result of §1.

Finally, in §5, we extend the results of §§1 and 3 to the case that external forces like friction or gyroscopic terms are present.

§1. The Homogenization Result

Loosely speaking, the homogenization result states the following. Provided the potential U is "nice" and certain resonance conditions are fulfilled, the limit x_0 describes the dynamics of a natural mechanical system on the submanifold N. The potential of this system can explicitly be constructed from U and the given initial values.

The precise statement given in §1.5 requires the introduction of some notion first. The discussion of the model problem in §I.2 can serve as a motivation for most of the definitions.

§1.1. Natural Mechanical Systems with a Strong Potential

Let M be a smooth[12] m-dimensional Riemannian manifold with metric $\langle \cdot, \cdot \rangle$. For a sequence $\epsilon \to 0$, we consider a family of mechanical systems on the configuration space M given by the Lagrangians

$$\mathscr{L}_\epsilon(x, \dot{x}) = \tfrac{1}{2}\langle \dot{x}, \dot{x}\rangle - V(x) - \epsilon^{-2}U(x), \qquad \dot{x} \in T_xM,$$

with smooth potentials V and U. We assume that V is bounded from below and U is non-negative.[13] The corresponding singularly perturbed equation of motion is given by the Euler-Lagrange equation, [1, Prop. 3.7.4],

$$\nabla_{\dot{x}_\epsilon}\dot{x}_\epsilon + \operatorname{grad} V(x_\epsilon) + \epsilon^{-2}\operatorname{grad} U(x_\epsilon) = 0, \tag{II.1}$$

where the *covariant derivative* ∇ denotes the Levi-Cività connection of the Riemannian manifold M. The *energy*, [1, Sect. 3.7],

$$E_\epsilon = \tfrac{1}{2}\langle \dot{x}_\epsilon, \dot{x}_\epsilon\rangle + V(x_\epsilon) + \epsilon^{-2}U(x_\epsilon),$$

is a constant of motion. If we assume that the energy surfaces $E_\epsilon = \text{const}$ are compact submanifolds of the tangent bundle TM, the flow of the equation of motion (II.1) is *complete*, [1, Prop. 2.1.17]. This means that any corresponding initial value problem is solvable *for all times*.

---

[12]The term "smooth" will mean "at least four times continuously differentiable" throughout this chapter.

[13]For potentials other than this there is *no* reasonable singular limit. For instance, $U(x) = -x^2/2$ yields (setting $V = 0$) an exponentially diverging family $x_\epsilon = \epsilon \sinh(t/\epsilon)$ of solutions with energy $E_\epsilon = 1/2$, independent of ϵ.

Since we study the singular limit behavior of a *family* of mechanical problems it is physically reasonable to bound the energy uniformly in ϵ, $E_\epsilon \leq E_*$. In fact, this is a condition on the initial values, which we choose to be fixed in the positions and converging in the velocities,[14]

$$x_\epsilon(0) = x_*, \qquad \lim_{\epsilon \to 0} \dot{x}_\epsilon(0) = v_* \in T_{x_*}M. \qquad (\text{II.2})$$

The equi-boundedness of the energy directly implies that $U(x_*) = 0$. Therefore, energy itself converges as a number in $\mathbb{R}$,

$$E_\epsilon \to E_0 = \tfrac{1}{2}\langle v_*, v_* \rangle + V(x_*). \qquad (\text{II.3})$$

§1.2. The Critical Submanifold

The set where the potential U vanishes is of utmost importance for the limit behavior under study.

Definition 1. Let the potential U be non-negative, $U \geq 0$, and let $N = \{x \in M : U(x) = 0\} \subset M$ be a compact,[15] smoothly embedded n-dimensional submanifold such that $N = \{x \in M : DU(x) = 0\}$ and the Hessian H of U, defined as a field of linear operators $H : TM|N \to TM|N$ by[16]

$$\langle H(x)u, v \rangle = D^2U(x)(u,v), \qquad u, v \in T_xM, \; x \in N,$$

fulfills the nondegeneracy condition

$$\ker H(x) = T_xN, \qquad x \in N. \qquad (\text{II.4})$$

Then N will be called a *nondegenerate critical manifold*[17] of U and the potential U will be called *constraining to N*.

There are two equivalent formulations of the nondegeneracy condition (II.4) that we will frequently use. As the Hessian H is selfadjoint with respect to the Riemannian metric, a first equivalent is given by

$$\operatorname{range} H(x) = T_xN^\perp, \qquad x \in N. \qquad (\text{II.5})$$

---

[14]This particular choice is for simplicity and elegance of the result only. At the expense of a far more technical result, one could consider converging initial positions as well.

[15]There is no loss of generality since we deal with *compact* energy surfaces and finite time intervals only.

[16]The Hessian H is *invariantly* defined on the tangent bundle *restricted* to base-points in N only. Note, that the second derivative D^2U is *not* a tensor field but coordinate dependent in general,

$$\frac{\partial^2 U}{\partial x^i \partial x^j} = \frac{\partial^2 U}{\partial \bar{x}^k \partial \bar{x}^l} \frac{\partial \bar{x}^k}{\partial x^i} \frac{\partial \bar{x}^l}{\partial x^j} + \frac{\partial U}{\partial \bar{x}^k} \frac{\partial^2 \bar{x}^k}{\partial x^i \partial x^j}.$$

However, the second term of the right hand side vanishes on the *critical* manifold N.

[17]This notion was introduced by BOTT [20] in his study of parameter-dependent Morse theory.

Using the fact that $U \geq 0$ and a compactness argument, we obtain as a second equivalent that there is a constant $\omega_* > 0$ such that

$$\langle H(x)u, u \rangle \geq \omega_*^2 \langle u, u \rangle, \qquad u \in T_x N^\perp, \quad x \in N. \tag{II.6}$$

Parts of these conditions can be described conveniently using certain *projections*. Let $P : TM|N \to TN^\perp$ and $Q : TM|N \to TN$ be the bundle maps defined by letting

$$P(x) : T_x M \to T_x N^\perp, \qquad Q(x) : T_x M \to T_x N, \qquad x \in N,$$

the orthogonal projections of $T_x M$ onto $T_x N^\perp$ and onto $T_x N$, respectively. Now, the nondegeneracy conditions imply

$$PH = H, \qquad QH = 0. \tag{II.7}$$

§1.3.   Spectrally Smooth Constraining Potentials

We now introduce the class of constraining potentials U for which the homogenization problem is solvable. For each $x \in N$ the Hessian $H(x)$ is a selfadjoint linear operator on $T_x M$. Therefore, it is diagonizable. We will need that the spectrum can be arranged in a contiguous way.

Definition 2. Let U be a potential constraining to the n-dimensional nondegenerate critical submanifold N. If the Hessian H of U has a smooth spectral decomposition on N,

$$H(x) = \sum_{\lambda=1}^{s} \omega_\lambda^2(x) P_\lambda(x), \qquad x \in N, \tag{II.8}$$

the potential U will be called to constrain *spectrally smooth* to N. Here, the smooth bundle maps $P_\lambda : TM|N \to TN^\perp$ define by $P_\lambda(x) : T_x M \to T_x N^\perp$ ($x \in N$) orthogonal projections of $T_x M$ onto mutually orthogonal subspaces of $T_x N^\perp$.

The nondegeneracy condition (II.5) implies that $P = \sum_\lambda P_\lambda$. For reference, we state the orthogonality properties of these projections explicitly,

$$P_\lambda^2 = P_\lambda, \qquad P_\lambda P_\mu = 0 \quad \lambda \neq \mu, \qquad P_\lambda^* = P_\lambda, \tag{II.9}$$

denoting by P_λ^* the adjoint linear operator with respect to the Riemannian metric.

The smooth scalar fields $\omega_\lambda^2 : M \to \mathbb{R}$ represent all the *nonzero* eigenvalues of the Hessian; the nondegeneracy condition (II.6) is equivalent to the uniform lower bound

$$\omega_\lambda(x) \geq \omega_* > 0, \qquad x \in N. \tag{II.10}$$

Therefore, the square root ω_λ of each eigenvalue constitutes a smooth function. For reasons which will become clear later on, we call these functions the *normal frequencies* of H. Without loss of generality we may assume that they are mutually non-identical on N since otherwise we could combine the corresponding eigenprojections. Moreover, the integers

$$n_\lambda = \operatorname{tr} P_\lambda = \dim \operatorname{range} P_\lambda \in \mathbb{N}, \qquad \sum_\lambda n_\lambda = r = m - n,$$

are *constants* on N, summing up to the *codimension* $r = m - n$ of the critical submanifold N. We call n_λ the *smooth multiplicity* of the y-dependent family ω_λ^2 of eigenvalues. Notice, that the multiplicity of $\omega_\lambda^2(y)$ at a particular point $y \in N$ might be accidentally greater than n_λ. Such points are called *resonance* points.

§1.4. Resonance Conditions

The homogenization result we are going to prove relies on imposing certain resonance conditions on the normal frequencies ω_λ. A resonance of order $j \in \mathbb{N}$ is given by the relation

$$\gamma_1 \omega_1 + \ldots + \gamma_s \omega_s = 0, \qquad |\gamma_1| + \ldots + |\gamma_s| = j, \qquad \text{(II.11)}$$

with integer coefficients $\gamma_\lambda \in \mathbb{Z}$. The nondegeneracy condition (II.10) implies that there are *no* resonances of order *one*. In general, each resonance relation (II.11) constitutes a *hypersurface* in the critical submanifold N.

Definition 3. Let $k \in \mathbb{N}$ be given. Assume that a time-dependent trajectory x on the critical submanifold N crosses all hypersurfaces of resonances of orders $j \leq k$ only transversally.[18] Then, x will be called *non-flatly resonant up to order k*.

§1.5. The Statement of the Homogenization Result

We are going to state that the limit dynamics is given by a natural mechanical system which involves a certain new potential. This potential can be constructed from the constraining potential U and the initial values.

Definition 4. Let U be a potential constraining spectrally smooth to the manifold N. Introducing the constants

$$\theta_*^\lambda = \frac{\langle P_\lambda(x_*)v_*, P_\lambda(x_*)v_*\rangle}{2\,\omega_\lambda(x_*)}, \qquad \lambda = 1, \ldots, s, \qquad \text{(II.12)}$$

---

[18]More precisely, since the resonance sets under consideration might be geometrically degenerate and not form hypersurfaces, we only assume that

$$\frac{d}{dt}\left(\gamma_1 \omega_1(x(t_r)) + \ldots \gamma_s \omega_s(x(t_r))\right) \neq 0$$

for all impact times t_r such that the resonance (II.11) holds at $x(t_r)$.

we set

$$U_{\text{hom}}(x) = \sum_{\lambda=1}^{s} \theta_*^{\lambda} \omega_{\lambda}(x), \qquad x \in N. \tag{II.13}$$

The potential U_{hom} will be called the *homogenization* of the constraining potential U with respect to the initial values $v_* \in T_{x_*}M$, $x_* \in N$.

Physically speaking, the constants θ_*^{λ} have the dimension of an *action* as a ratio of energy and frequency. Later, in §2, our proof will identify them as the *adiabatic invariants* of the motion normal to the constraint manifold N.

Theorem 1. *For a sequence $\epsilon \to 0$, consider the family of mechanical systems given by the Lagrangian*

$$\mathscr{L}_{\epsilon}(x, \dot{x}) = \tfrac{1}{2}\langle \dot{x}, \dot{x} \rangle - V(x) - \epsilon^{-2}U(x), \qquad \dot{x} \in T_xM.$$

The potential U is assumed to constrain spectrally smooth to a nondegenerate critical submanifold $N \subset M$. Let the initial positions be fixed on the critical submanifold, $x_{\epsilon}(0) = x_* \in N$, and the initial velocities convergent in $T_{x_*}M$, $\dot{x}_{\epsilon}(0) \to v_* \in T_{x_*}M$. Then, for a finite time interval $[0, T]$, there exists a unique sequence x_{ϵ} of solutions of the Euler-Lagrange equations corresponding to $\mathscr{L}_{\epsilon}$.

Let U_{hom} be the homogenization of U with respect to the limit initial values (x_*, v_*). We denote by x_{hom} the unique solution of the Euler-Lagrange equations corresponding to the homogenized Lagrangian

$$\mathscr{L}_{\text{hom}}(x, \dot{x}) = \tfrac{1}{2}\langle \dot{x}, \dot{x} \rangle - V(x) - U_{\text{hom}}(x), \qquad \dot{x} \in T_xN,$$

with initial data $x_{\text{hom}}(0) = x_* \in N$ and $\dot{x}_{\text{hom}}(0) = Q(x_*)v_* \in T_{x_*}N$.

If x_{hom} is non-flatly resonant up to order three, the sequence x_{ϵ} converges uniformly to x_{hom} on $[0, T]$.

§1.6. Homogenization of a Specific Class of Potentials

In applications, one frequently encounters constraining potentials of the form

$$U(x) = \tfrac{1}{2}\sum_{j=1}^{r} |\psi_j(x)|^2, \qquad x \in M,$$

with smooth scalar functions $\psi_j : M \to \mathbb{R}$. This way, the critical submanifold N is given by the set

$$N = \{x \in M : U(x) = 0\} = \{x \in M : \psi_1(x) = \ldots = \psi_r(x) = 0\}.$$

This manifold N has codimension r if and only if the gradient vectors, $\text{grad}\,\psi_j(x)$, are linearly independent for all $x \in N$, which we will assume to be the case. Then, they form a basis field for the normal bundle of N,

$$\dim T_x N^{\perp} = r, \quad \mathrm{span}\{\mathrm{grad}\,\psi_j(x) : j = 1,\ldots,r\} = T_x N^{\perp}, \quad x \in N. \tag{II.14}$$

In coordinates, the second derivative of U is given by

$$D^2 U(x) = \sum_j D\psi_j(x)^T D\psi_j(x), \quad x \in N.$$

Recalling that $\mathrm{grad}\,\psi_j = G^{-1} D\psi_j^T$, cf. [1, Def. 2.5.14], where G denotes the metric tensor of the coordinate system, we get the following expression for the Hessian H:

$$H = G^{-1} D^2 U\big|_N = \sum_j \langle\,\cdot\,, \mathrm{grad}\,\psi_j\rangle\,\mathrm{grad}\,\psi_j.$$

Notice, that the last expression is independent of a chosen coordinate system. Using the span relation (II.14), we obtain the nondegeneracy condition (II.5), range $H = T^{\perp} N$. Thus, U constrains to the submanifold N in the technical sense of Definition 1.

Now, let $\omega^2(x) > 0$ be a nonzero eigenvalue of $H(x)$ and $X \in T_x N^{\perp}$ the corresponding eigenvector,

$$H(x)X = \omega^2(x)X. \tag{II.15}$$

Because of (II.14), there is a unique representation of X of the form

$$X = \sum_j \xi^j \cdot \mathrm{grad}\,\psi_j(x).$$

Inserting this into the eigenvalue problem (II.15) yields an equivalent, r-dimensional problem,

$$H_r(x)\xi = \omega^2(x)\xi, \tag{II.16}$$

for the *Grammian matrix* of the gradient vectors,

$$H_r(x)_{jk} = \langle \mathrm{grad}\,\psi_j(x), \mathrm{grad}\,\psi_k(x)\rangle. \tag{II.17}$$

The matrix-valued field $H_r : N \to \mathbb{R}^{r \times r}$ will be called the *reduced Hessian* of the constraining potential U. Notice, that the normalization of X is given according to

$$\langle X, X\rangle = \xi^T H_r(x)\xi.$$

Solving the reduced $r \times r$ eigenvalue problem (II.16) is all one needs to know for, first, deciding about whether U constrains *spectrally* smooth, and second, establishing the homogenized potential U_{hom}.

Example 1. (The Codimension One Case). The case of codimension $r = 1$ provides the simplest possible example. Suppressing the index '1,' we directly read off that

$$\omega(x) = \| \operatorname{grad} \psi(x) \|, \qquad x \in N,$$

where $\| \cdot \|$ denotes the norm on the tangent space $T_x M$ induced by the Riemannian metric. The adiabatic invariant θ_* is given by

$$\theta_* = \frac{\langle v_*, \operatorname{grad} \psi(x_*) \rangle^2}{2\,\omega^3(x_*)}.$$

Finally, U_{hom} takes the form $U_{\text{hom}}(x) = \theta_* \cdot \omega(x)$.

§1.7.   Remarks on Genericity

We now study the "genericity" of the assumption that U constrains *spectrally smooth*. By definition, *generic* properties are structurally stable under perturbations, i.e., are typical for a whole neighborhood of problems. Otherwise, a property would be highly improbable in the presence of some uncertainty in the problem data like modeling or measurement errors.

The reader should be cautioned that the term "generic" is always understood relatively to a *specific* class of perturbations.

Now, the genericity of the spectral smoothness of U depends on two different influences: first, the presence of resonances on N; second, the class of perturbations of U allowed for by the underlying physical problem.

The Nonresonant Case.   If there are no resonances of order two of the normal frequencies ω_λ on N, the potential U constrains spectrally smooth. This claim can be proved by representing the spectral projections by a Cauchy integral, a standard argument of perturbation theory [51]. We omit the technical details here, the reader will find them in the first part of the proof of Theorem II.3 later on.

Accordingly, because having no resonances of order two is certainly a generic property with respect to *any* class of perturbations, the potential U constrains spectrally smooth *generically*.

The Resonant Case.   Here, certain resonances of order two exist on the manifold N, a far more subtle case which requires a careful study of the perturbations that are allowed. These perturbations determine the so-called *codimension*[19] of a resonance, i.e., the codimension of the set of all matrices having that eigenvalue-resonance in the set of all those matrices that are consistent with the given class of perturbations.

For real symmetric matrices, as the Hessian H, we typically have resonances of codimension one or two, depending on whether the perturbations

---

[19]This notion should not be confused with the codimension r of the manifold N.

preserve certain structures besides the symmetry $H = H^T$ or not. For a complete classification of generic resonances in a problem class related to quantum mechanics we refer to the remarks given in §IV.2.4.

Instead of developing a general theory we just provide some examples and remarks which, however, seem to reflect the general situation quite well.

Example 2. Let the constraining potential U have the special structure

$$U(x) = \tfrac{1}{2} \sum_{j=1}^{r} |\psi_j(x_j)|^2, \qquad x \in M.$$

Suppose that the underlying problem only admits perturbations which preserve this structure. The results of §1.6 show that the reduced Hessian H_r is always diagonal for this specific class of potentials, implying that U is *generically* spectrally smooth. Now, the set of resonant diagonal matrices obviously has *codimension one* in the set of all diagonal matrices.

The discussion of the butane molecule in §III.2.1 will provide a less trivial example where a potential is generically spectrally smooth because just a restricted class of perturbations applies.

The most general class of perturbations of a potential leads to perturbations of the Hessian that preserve the symmetry of the matrix only. Here, a *generic* resonance has codimension *two* and, unavoidably, the potential does *not* constrain spectrally smooth. A proof of these claims can be found in Appendix A. *If we have to admit general perturbations of the constraining potential, either a resonance or the spectral smoothness are non-generic.*

An example of a generic resonance of codimension two is subject of §4. We will show that the convergence assertion of Theorem 1, which is necessarily not applicable there, suffers a dramatic breakdown.

§1.8.   A Counterexample for Flat Resonances

The short discussion in §I.2.6 has motivated why resonance conditions have to be employed at a certain stage of the *proof* we designed for Theorem 1. Here, we will show that the *result* itself demands resonance conditions.

We consider the Euclidean space $M = \mathbb{R}^3$ and the potential

$$U(x) = \tfrac{1}{2}\omega_1^2(y)|z^1|^2 + \tfrac{1}{2}\omega_2^2(y)|z^2|^2, \qquad x = (y, z^1, z^2) \in \mathbb{R}^3,$$

with smooth functions $\omega_1, \omega_2 \geq 1$ which will be specified below. This potential is obviously constraining—in the technical sense of Definition 1— to the one-dimensional submanifold $N = \{x \in M : z^1 = z^2 = 0\}$. The Hessian H is given by the diagonal matrix

$$H(y) = \mathrm{diag}(0, \omega_1^2(y), \omega_2^2(y)).$$

The initial values shall be given by $x_* = 0 \in N$ and $v_* = (w_*, 0, 2)^T$. Now, we specify ω_1 and ω_2. Let $\omega \in C^\infty(\mathbb{R})$ be a function such that

$$\omega(y) = \begin{cases} 1, & y \leq 1/2, \\ 0, & 1 \leq y \leq 2, \\ 1, & 5/2 \leq y, \end{cases}$$

and $\omega'(y) \neq 0$ for $1/2 < y < 1$ and $2 < y < 5/2$. We set

$$\omega_1 = 1, \qquad \omega_2 = 1 + \omega.$$

Defining the projections $P_1 = \mathrm{diag}(0, 1, 0)$ and $P_2 = \mathrm{diag}(0, 0, 1)$, the Hessian H trivially has the smooth spectral decomposition

$$H(y) = \omega_1^2(y)P_1 + \omega_2^2(y)P_2. \tag{II.18}$$

According to Definition 4, this and the initial values yield the homogenized potential

$$U_{\mathrm{hom}} = \omega_2.$$

However, because of the resonance there are other smooth spectral decompositions of H which follow the paths of the eigenvalues in a different way. Let $\phi \in C^\infty(\mathbb{R})$ be a function such that

$$\phi(y) = \begin{cases} 0, & y \leq 1, \\ \pi/2, & 2 \leq y, \end{cases}$$

and define the mutually orthogonal projections

$$\hat{P}_1(\phi) = \begin{pmatrix} 0 & 0 & 0 \\ 0 & \cos^2\phi & \cos\phi\sin\phi \\ 0 & \cos\phi\sin\phi & \sin^2\phi \end{pmatrix},$$

and

$$\hat{P}_2(\phi) = \begin{pmatrix} 0 & 0 & 0 \\ 0 & \sin^2\phi & -\cos\phi\sin\phi \\ 0 & -\cos\phi\sin\phi & \cos^2\phi \end{pmatrix}.$$

In particular, we have $\hat{P}_1(0) = P_1$ and $\hat{P}_2(0) = P_2$, but $\hat{P}_1(\pi/2) = P_2$ and $\hat{P}_2(\pi/2) = P_1$. A short calculation reveals the smooth spectral decomposition

$$H(y) = \hat{\omega}_1^2(y)\hat{P}_1(\phi(y)) + \hat{\omega}_2^2(y)\hat{P}_2(\phi(y)), \tag{II.19}$$

where

$$\hat{\omega}_1(y) = \begin{cases} \omega_1(y), & y \leq 3/2, \\ \omega_2(y), & 3/2 \leq y, \end{cases} \qquad \hat{\omega}_2(y) = \begin{cases} \omega_2(y), & y \leq 3/2, \\ \omega_1(y), & 3/2 \leq y. \end{cases}$$

Correspondingly, Definition 4 yields the homogenized potential

$$\hat{U}_{\text{hom}} = \hat{\omega}_2.$$

The force fields U'_{hom} and $\hat{U}'_{\text{hom}}$ differ, since for $y \in]2, 5/2[$ we have

$$U'_{\text{hom}}(y) = \omega'_2(y) = \omega'(y) \neq 0 = \omega'_1(y) = \hat{U}'_{\text{hom}}(y).$$

For an initial velocity $w_* > 0$ which is large enough, the corresponding solutions x_{hom} and $\hat{x}_{\text{hom}}$ will hit this interval at some time $t < T = 1$. Therefore, they must be different, $x_{\text{hom}} \neq \hat{x}_{\text{hom}}$.

Clearly now, Theorem 1 cannot hold true after dropping the assumption of non-flat resonance. For then we would get the contradiction that both $y_\epsilon \to x_{\text{hom}}$ and $y_\epsilon \to \hat{x}_{\text{hom}}$, i.e., $x_{\text{hom}} = \hat{x}_{\text{hom}}$ on the time interval $[0, 1]$.

The setting of this counterexample fits into the framework of the model problem of §I.2. Therefore, Theorem I.1 is applicable. It teaches that the homogenized potential $U_{\text{hom}} = \omega_2$ is, in fact, the "correct" one, describing the limit dynamics and yielding the uniform convergence $y_\epsilon \to x_{\text{hom}}$. Thus, the spectral decomposition (II.18) of the Hessian is somewhat more "natural" than the other, constructed one, Eq. (II.19). However, it seems to be difficult defining a general notion of "natural" smooth spectral decompositions in a way which would allow to relax the resonance conditions of Theorem 1 in any significant fashion. Moreover, the author conjectures that the resonance conditions of Theorem 1 do not only reflect the possibility of a non-uniqueness of the homogenized potential, but also the possibility of a complete breakdown of the entire limit structure.

§1.9. A Counterexample for Unbounded Energy

If we drop the assumption of uniformly bounded energy, i.e., the assumption $x_* \in N$, we cannot expect a homogenization result similar to Theorem 1. To show this, we reproduce a counterexample of BORNEMANN and SCHÜTTE [18].

We consider on the configuration space $M = \mathbb{R}$ the Lagrangian

$$\mathscr{L}_\epsilon(x, \dot{x}) = \tfrac{1}{2}\dot{x}^2 - \epsilon^{-2}U(x)$$

with the potential[20]

$$U(x) = \begin{cases} x^2/2 & x \leq 0, \\ 2x^2 & x \geq 0, \end{cases}$$

which is constraining to the manifold $N = \{0\}$. Given the initial position $x_* = 1 \notin N$ and the initial velocity $v_* = 0$, the energy $E_\epsilon = 2\epsilon^{-2} \to \infty$

---

[20]The limited differentiability of U is *not* essential for this counterexample. It could be smoothed out at the cost of sacrificing the simplicity of the result.

cannot be bounded uniformly in ϵ. The solution of the equation of motion is given by the rapidly oscillating function $x_\epsilon(t) = x(t/\epsilon)$, where

$$x(t) = \begin{cases} \cos(2t) & 0 \le t \le \pi/4, \\ -2\sin(t - \pi/4) & \pi/4 \le t \le 5\pi/4, \\ \sin(2t - 5\pi/2) & 5\pi/4 \le t \le 3\pi/2. \end{cases}$$

Here, we get merely *weak** convergence of x_ϵ in L^∞, namely by a well-known generalization, [23, Lemma I.1.2], of the Riemann-Lebesgue lemma

$$x_\epsilon \overset{*}{\rightharpoonup} x_0 \equiv -2/\pi = \frac{1}{3\pi/2} \int_0^{3\pi/2} x(\tau)\, d\tau,$$

which is *not* on the constraint manifold N. Trivially, this limit cannot be described by a mechanical system with configuration space N.

§1.10. Bibliographical Remarks

Strikingly, the homogenization problem of this chapter has found only little systematic attention in the mathematical and physical literature—at least much less than the special case provided by the realization-of-constraints-problem that will be discussed in §3.

All work we know of about this particular homogenization problem considers an Euclidean space $M = \mathbb{R}^d$ as the configuration manifold. The generalization to Riemannian manifolds, as accomplished by us, is *not* straightforward since the metric introduces a further source of nonlinearity. As we will see, the difficulty of the proof consists in controlling the nonlinearities with respect to weak* convergence.

To our knowledge, the first mathematical work on the homogenization problem was done by RUBIN and UNGAR [82, p. 82f.] in 1957. These authors consider constraint manifolds N of codimension $r = 1$ only. The result, somewhat hidden in the paper, is stated in a special coordinate system.

Independently, for codimension $r = 1$, the result can be found by means of an example in the work of the physicists KOPPE and JENSEN [58, Eq. (7)] from 1971. The argument of these authors is basically physical and involves averaging in an informal way. However, we owe the backbone of our proof to the physical ideas presented in their paper: a virial theorem and adiabaticity. The notion of weak* convergence puts the informal averaging process on a firm mathematical basis. A corresponding mathematical proof of the codimension $r = 1$ case has been worked out by BORNEMANN and SCHÜTTE [18].

The codimension $r = 1$ case was also discussed in the work of the physicist VAN KAMPEN [49] in 1985. This author utilizes the WKB method. However, the proof is mathematically incomplete as has been pointed out by BORNEMANN and SCHÜTTE [18].

The first—and until our present work only—complete study of the general case for $M = \mathbb{R}^d$ was given by TAKENS [94] in 1979. This author revealed the importance of resonance conditions and proved the result under the assumptions that there are *no* resonance hypersurfaces of second and third order, all eigenvalues of the Hessian having smooth multiplicity *one*. The method of proof set up by TAKENS starts with a rather explicit representation of the normal oscillations, as one would do for an asymptotic analysis, and then proceeds by using the Riemann-Lebesgue lemma, which is in fact a result about weak convergence. The presentation of the proof is in parts only sketchy, cf. Remark 2 on p. 43 below.

Referring to the work of TAKENS, an informal discussion of the general homogenization result can be found in an article by KOILLER [57] from 1990. This author uses action-angle variables and identifies the constants θ_*^λ in the expression for the homogenized potential U_{hom} as *adiabatic invariants*.

One should also mention the work of KELLER and RUBINSTEIN [52] from 1991. The concern of these authors is the homogenization problem for the semilinear wave equation $v_{tt} = \Delta v - \epsilon^{-2} \operatorname{grad} U(v)$. Their argument—an ingenious multiple scale asymptotics—is directly applicable to our problem and correctly establishes the homogenization potential U_{hom}. However, the expansion presented in the paper is only *formal* and no estimate of the remainder term is given. These authors cannot predict difficulties at resonances like in §1.8. It should be stressed that no approximation further than the zero-order term x_0, the singular limit, is provided.

§2.   The Proof of the Homogenization Result

The proof of Theorem 1 proceeds along the lines we have discussed in §I.2. However, in contrast to the simple example given there, we now need a considerable amount of notions from differential geometry. For purposes of reference, we collect some of the intermediate results in a series of sixteen lemmas. The assumptions of Theorem 1 shall be valid throughout.

§2.1.   Step 1: Equi-Boundedness (Energy Principle)

The proof will be given in *local coordinates*. Since we work on compact time intervals $[0, T]$ we can restrict ourselves to a single coordinate patch of M, where we have

$$x = (x^1, \ldots, x^m) \in \Omega \subset \mathbb{R}^m.$$

In these coordinates the metric is represented as usual by a covariant tensor of second order, $G = (g_{ij})$. The equation of motion (II.1) can be rewritten[21] as

$$\ddot{x}_\epsilon + \Gamma(\dot{x}_\epsilon, \dot{x}_\epsilon) + F_V(x_\epsilon) + \epsilon^{-2} F_U(x_\epsilon) = 0. \tag{II.20}$$

---

[21] Cf. [1, Proof of Prop. 3.7.4].

The forces $F_V = \operatorname{grad} V$ and $F_U = \operatorname{grad} U$ are given by[22]

$$F_V^i = g^{ij}\frac{\partial V}{\partial x^j}, \qquad F_U^i = g^{ij}\frac{\partial U}{\partial x^j},$$

where the contravariant tensor of second order $(g^{ij}) = G^{-1}$ represents the inverse matrix of G. The term Γ denotes the *Christoffel symbols*

$$\Gamma(u,v)^i = \Gamma_{jk}^i\, u^j v^k, \qquad \Gamma_{jk}^i = \tfrac{1}{2}g^{il}\left(\frac{\partial g_{jl}}{\partial x^k} + \frac{\partial g_{lk}}{\partial x^j} - \frac{\partial g_{jk}}{\partial x^l}\right).$$

We will frequently use a slightly nonsymmetric version of the Christoffel symbols,

$$\hat{\Gamma}_{jk}^i = g^{il}\left(\frac{\partial g_{jl}}{\partial x^k} - \frac{1}{2}\frac{\partial g_{jk}}{\partial x^l}\right),$$

which obviously gives $\Gamma(u,u) = \hat{\Gamma}(u,u)$. In coordinate representation the Hessian is given by the tensor

$$H = (H_j^i), \qquad H_j^i = g^{ik}\frac{\partial^2 U}{\partial x^k \partial x^j}$$

and the energy as the expression

$$E_\epsilon = \tfrac{1}{2}g_{ij}(x_\epsilon)\dot{x}_\epsilon^i\dot{x}_\epsilon^j + V(x_\epsilon) + \epsilon^{-2}U(x_\epsilon).$$

The nondegenerate critical submanifold N can now be viewed as a submanifold of $\mathbb{R}^m$.

Lemma 1. *There is a subsequence of $\epsilon \to 0$, denoted by ϵ again, such that*

$$x_\epsilon \to x_0 \quad \text{in} \quad C^0([0,T],\mathbb{R}^m), \qquad \dot{x}_\epsilon \overset{*}{\rightharpoonup} \dot{x}_0 \quad \text{in} \quad L^\infty([0,T],\mathbb{R}^m).$$

The limit function is at least Lipschitz continuous and takes values in the submanifold N, $x_0 \in C^{0,1}([0,T],N)$.

Proof. Let $\alpha > 0$ denote the smallest eigenvalue of the metric G, V_* a lower bound of the potential V, and E_* a uniform bound for the energy E_ϵ. Conservation of energy gives

$$\alpha|\dot{x}_\epsilon(t)|^2/2 \le E_* - V_*$$

and integration

$$|x_\epsilon(t)| \le |x_*| + T\sqrt{\frac{2}{\alpha}(E_* - V_*)}$$

for all $t \in [0,T]$. These equi-boundedness results allow the application of the extended Arzelà-Ascoli theorem, Principle I.4: There is a subsequence

---

[22]Throughout this chapter we will apply Einstein's summation convention.

of ϵ, denoted by ϵ again, and a limit function $x^0 \in C^{0,1}([0,T], \mathbb{R}^m)$ such that the asserted limit relations hold.

Multiplying the equation of motion (II.20) by ϵ^2 and taking the weak* limit shows that

$$\operatorname{grad} U(x_0(t)) = 0, \quad \text{i.e.,} \qquad x_0(t) \in N,$$

for all $t \in [0, T]$. Here, we have used Principle I.1 for establishing the weak* convergence $\epsilon^2 \ddot{x}_\epsilon \overset{*}{\rightharpoonup} 0$. □

From now on we consider the subsequence of this lemma. Further extractions will follow and they will always be denoted by ϵ again.

The uniform convergence $x_\epsilon \to x_0$ implies that for sufficiently small ϵ all trajectories are within a *tubular neighborhood* of N. A point $x \in M$ is an element of such a neighborhood, if there is a unique representation

$$x = \exp_y z, \qquad y \in N, \quad z \in T_y N^\perp. \tag{II.21}$$

Here, "exp" denotes the geodesic exponential map, [1, p. 149], and a well-known theorem of differential geometry states the existence of such a tubular neighborhood of N, [1, Theorem 2.7.5]. We consider a smooth field $(e_1, \ldots, e_r)$ of orthonormal bases of the normal bundle $TN^\perp$, i.e., for $y \in N$

$$e_i(y) \in T_y N^\perp, \qquad \langle e_i(y), e_j(y) \rangle = \delta_{ij}.$$

Thus, in the tubular neighborhood we get the unique representation[23]

$$x = \exp_y \left(z^{n+1} e_1(y) + \ldots + z^{n+r} e_r(y) \right).$$

Well-known properties of geodesics, [1, p. 150], imply that

$$\operatorname{dist}(x, N)^2 = \langle z, z \rangle = \sum_{i=n+1}^{n+r} (z^i)^2, \tag{II.22}$$

where the left hand side is independent of the chosen bases field.

For a set of given local coordinates $(y^1, \ldots, y^n)$ of the manifold N we define the *tubular coordinates*

$$(x^1, \ldots, x^m) = (y^1, \ldots, y^n; z^{n+1}, \ldots, z^{n+r}).$$

Putting $y = (y^1, \ldots, y^n; 0, \ldots, 0)$ and $z = (0, \ldots, 0; z^{n+1}, \ldots, z^{n+r})$, this coordinate system has an obvious linear structure and we will frequently write in short form[24]

$$x = y + z, \qquad y \in N, \quad z \in T_y N^\perp.$$

---

[23]This particular numbering of the coefficients z^i will simplify the notation following below.

[24]Certainly, this is "abus de langage", compared to the invariant relation (II.21). However, when we use this short form, it should be clear that we are working in this specific coordinate system.

In the following we consider ϵ sufficiently small such that

$$x_\epsilon = \exp_{y_\epsilon} z_\epsilon, \text{ resp.}, \qquad x_\epsilon = y_\epsilon + z_\epsilon,$$

defines a time-dependent functions y_ϵ with values in N and z_ϵ with values in $T_{y_\epsilon} N^\perp$. We call y_ϵ the *constrained motion* of x_ϵ and z_ϵ its *normal motion*.

Lemma 2. *The limit relations specify as follows:*

$$y_\epsilon \to x_0 \quad \text{in} \quad C^0([0,T], N), \qquad \dot{y}_\epsilon \overset{*}{\rightharpoonup} \dot{x}_0 \quad \text{in} \quad L^\infty([0,T], \mathbb{R}^n),$$

for the constrained motion and[25]

$$z_\epsilon = O(\epsilon), \qquad \dot{z}_\epsilon \overset{*}{\rightharpoonup} 0 \quad \text{in} \quad L^\infty([0,T], \mathbb{R}^r)$$

for the normal motion.

Proof. Lemma 1 and the distance relation (II.22) proves that

$$y_\epsilon \to x_0, \quad \dot{y}_\epsilon \overset{*}{\rightharpoonup} \dot{x}_0, \quad z_\epsilon \to 0, \quad \dot{z}_\epsilon \overset{*}{\rightharpoonup} 0.$$

Conservation of energy and a Taylor expansion shows that

$$\epsilon^2 (E_* - V_*) \geq U(x_\epsilon) = \tfrac{1}{2} \langle H(y_\epsilon) z_\epsilon, z_\epsilon \rangle + O(|z_\epsilon|^3).$$

The nondegeneracy condition (II.6) and the uniform convergence $z_\epsilon \to 0$ imply, for sufficiently small ϵ, the estimate

$$|z_\epsilon|^2 \leq c(\epsilon^2 + |z_\epsilon|^3) \leq c\epsilon^2 + \tfrac{1}{2}|z_\epsilon|^2,$$

where c denotes some positive constant. This proves $z_\epsilon = O(\epsilon)$.   □

In what follows, we will always abbreviate $f_\epsilon = f(y_\epsilon)$ for any smooth tensor field defined on the submanifold N, including $\epsilon = 0$, i.e., $f_0 = f(x_0)$.

We now take a closer look on the constrained and normal motion. To this end, we apply the orthogonal projections of §1.2. Using the notation just introduced, we get

$$P_\epsilon z_\epsilon = z_\epsilon, \qquad P_\epsilon \dot{y}_\epsilon = 0.$$

Within the tubular coordinate system, we can view P_ϵ as a time dependent matrix. Thus, we can extend the action of P_ϵ in a coordinate dependent fashion and get for the velocities

$$P_\epsilon \dot{x}_\epsilon = P_\epsilon \dot{z}_\epsilon = \dot{z}_\epsilon - \dot{P}_\epsilon z_\epsilon = \dot{z}_\epsilon + O(\epsilon), \tag{II.23}$$

and for the accelerations

$$P_\epsilon \ddot{x}_\epsilon = \frac{d}{dt}(P_\epsilon \dot{x}_\epsilon) - \dot{P}_\epsilon \dot{x}_\epsilon = \ddot{z}_\epsilon - 2\dot{P}_\epsilon \dot{z}_\epsilon - \dot{P}_\epsilon \dot{y}_\epsilon + O(\epsilon) = \ddot{z}_\epsilon + O(1). \tag{II.24}$$

---

[25] Here and in what follows, we denote an estimate for a time-dependent function by a Landau symbol, if the estimate holds *uniformly* in time.

Lemma 3. *After a further extraction of subsequences the following limit relations are satisfied by the constrained motion:*

$$y_\epsilon \to x_0, \qquad \dot{y}_\epsilon \to \dot{x}_0, \qquad \ddot{y}_\epsilon \overset{*}{\to} \ddot{x}_0.$$

In particular, one gets the regularity $x_0 \in C^{1,1}([0,T], N)$. The initial values of x_0 are given by

$$x_0(0) = x_* \in N, \qquad \dot{x}_0(0) = Q(x_*)v_* \in T_{x_*}N.$$

The normal motion satisfies a second order equation of the form

$$\ddot{z}_\epsilon + \epsilon^{-2} H_\epsilon z_\epsilon = O(1). \tag{II.25}$$

Proof. If we multiply the equation of motion (II.20) by Q_ϵ, we get

$$\ddot{y}_\epsilon + (Q_\epsilon \ddot{x}_\epsilon - \ddot{y}_\epsilon) + Q_\epsilon \Gamma(x_\epsilon)(\dot{x}_\epsilon, \dot{x}_\epsilon) + Q_\epsilon F_V(x_\epsilon) + \epsilon^{-2} Q_\epsilon F_U(x_\epsilon) = 0.$$

The equi-boundedness of x_ϵ and $\dot{x}_\epsilon$ shows that $Q_\epsilon \Gamma(x_\epsilon)(\dot{x}_\epsilon, \dot{x}_\epsilon) = O(1)$ and $Q_\epsilon F_V(x_\epsilon) = O(1)$. The projection relation (II.24) yields $Q_\epsilon \ddot{x}_\epsilon - \ddot{y}_\epsilon = O(1)$. Below, Lemma 5 will state that $\epsilon^{-2} Q_\epsilon F_U(x_\epsilon) = O(1)$. Summarizing, we obtain

$$\ddot{y}_\epsilon = O(1).$$

An application of the extended Arzelà-Ascoli theorem, Principle I.4, proves the limit assertions.

Likewise, if we multiply the equation of motion (II.20) by P_ϵ, we get

$$\ddot{z}_\epsilon + (P_\epsilon \ddot{x}_\epsilon - \ddot{z}_\epsilon) + P_\epsilon \Gamma(x_\epsilon)(\dot{x}_\epsilon, \dot{x}_\epsilon) + P_\epsilon F_V(x_\epsilon) + \epsilon^{-2} P_\epsilon F_U(x_\epsilon) = 0.$$

Below, Lemma 5 will state that $\epsilon^{-2} P_\epsilon F_U(x_\epsilon) = \epsilon^{-2} H_\epsilon z_\epsilon + O(1)$. The same arguments as for the constrained motion yield that the middle terms are equi-bounded. Thus, we obtain the asserted second order equation.

Since $x_\epsilon(0) = x_*$ we obtain by the uniform convergence $x_\epsilon \to x_0$ that $x_0(0) = x_*$. From Eq. (II.23) follows

$$Q(x_*)v_* = \lim_{\epsilon \to 0} Q_\epsilon \dot{x}_\epsilon|_{t=0} = \lim_{\epsilon \to 0} \dot{y}_\epsilon(0).$$

The uniform convergence $\dot{y}_\epsilon \to \dot{x}_0$ implies $\dot{x}_0(0) = Q(x_*)v_*$. □

As a result, we now know that all quantities $f_\epsilon = f(y_\epsilon)$ are strongly convergent in $C^1[0,T]$.

Lemma 4. *In $L^\infty([0,T], \mathbb{R}^r)$, the bounded quantity $\epsilon^{-1} z_\epsilon$ converges weakly* to zero,*

$$\epsilon^{-1} z_\epsilon \overset{*}{\to} 0.$$

Proof. Lemma 2 shows that

$$\eta_\epsilon = z_\epsilon/\epsilon = O(1).$$

After an application of the Alaoglu theorem, Principle I.3, and an extraction of subsequences, we obtain that

$$\eta_\epsilon \overset{*}{\rightharpoonup} \eta_0$$

for some $\eta_0 \in L^\infty([0,T], \mathbb{R}^r)$. Multiplying the relation (II.25) by ϵ and taking weak* limits gives, by Principle I.1, $H(x_0)\eta_0 = 0$. Thus, the nondegeneracy condition (II.4) yields that

$$\eta_0 \in T_{x_0}N.$$

On the other hand, since $\eta_\epsilon \in T_{y_\epsilon}N^\perp$ implies $Q_\epsilon\eta_\epsilon = 0$, we get by taking weak* limits that $Q(x_0)\eta_0 = 0$, i.e.,

$$\eta_0 \in T_{x_0}N^\perp.$$

Hence, we obtain that $\eta_0 = 0$. Since this limit is *unique* we may discard the extraction of subsequences, recalling Principle I.5.   □

Despite the fact that $\dot{z}_\epsilon \overset{*}{\rightharpoonup} 0$ and $z_\epsilon/\epsilon \overset{*}{\rightharpoonup} 0$, we *cannot* expect that quadratic expressions of these quantities converge weakly* to zero. This lack of weak* sequential convergence constitutes the core of the homogenization result and, as we will see in §3.1, the general obstruction for realization of constraints. For this reason, we explicitly introduce the quadratic expressions[26]

$$\Sigma_\epsilon = \epsilon^{-2}z_\epsilon \otimes G_\epsilon z_\epsilon = \epsilon^{-2}z_\epsilon \otimes z_\epsilon G_\epsilon = (\Sigma_{\epsilon j}^{i}), \qquad \Sigma_{\epsilon j}^{i} = \epsilon^{-2}z_\epsilon^i z_\epsilon^k g_{kj}(y_\epsilon),$$

and

$$\Pi_\epsilon = \dot{z}_\epsilon \otimes G_\epsilon \dot{z}_\epsilon = \dot{z}_\epsilon \otimes \dot{z}_\epsilon G_\epsilon = (\Pi_{\epsilon j}^{i}), \qquad \Pi_{\epsilon j}^{i} = \dot{z}_\epsilon^i \dot{z}_\epsilon^k g_{kj}(y_\epsilon).$$

By Lemma 2, both quadratic expressions are uniformly bounded,

$$\Sigma_\epsilon = O(1), \qquad \Pi_\epsilon = O(1).$$

We may therefore assume—by an application of the Alaoglu theorem, Principle I.3, and after a further extraction of subsequences—that

$$\Sigma_\epsilon \overset{*}{\rightharpoonup} \Sigma_0, \qquad \Pi_\epsilon \overset{*}{\rightharpoonup} \Pi_0.$$

Later on we will see that in general $\Sigma_0 \neq 0$ and $\Pi_0 \neq 0$.[27] There is a noteworthy difference in the definitions of Σ_ϵ and Π_ϵ: Whereas Σ_ϵ *invariantly*

---

[26]The metric tensor G_ϵ is included to make the resulting matrices *selfadjoint* with respect to the Riemannian metric. This simplifies the calculations later on.

[27]To be specific, we will prove in §3.1, Lemma 17, that $\Sigma_0 = 0$, or $\Pi_0 = 0$, if and only if the limit initial velocity v_* is *tangential* to the critical submanifold N, $v_* \in T_{x_*}N$.

defines a tensor field along y_ϵ, covariant of order one and contravariant of order one, the definition of Π_ϵ *depends* on the choice of the coordinate system.

We finish this step of the proof by stating a Taylor expansion of the strong force term F_U. Parts of this statement have been used in the proof of Lemma 3. For convenience, we introduce some further notation: The quantity $\mathrm{grad}\, H$ denotes

$$(\mathrm{grad}\, H)^{ij}_k = g^{il}\frac{\partial H^j_k}{\partial x^l}.$$

For $A = (A^{jk})$ and $B = (B^k_j)$ we define the traces

$$(\hat\Gamma : A)^i = \hat\Gamma^i_{jk}A^{jk}, \qquad (\mathrm{grad}\, H : B)^i = (\mathrm{grad}\, H)^{ij}_k B^k_j.$$

Lemma 5. *The Taylor expansions of the force term $\epsilon^{-2}F_U$ to second order is given by*

$$\begin{aligned}
\epsilon^{-2}F_U(x_\epsilon) &= \epsilon^{-2}H_\epsilon z_\epsilon + O(1) \\
&= \epsilon^{-2}H_\epsilon z_\epsilon - \hat\Gamma_\epsilon : (H_\epsilon\Sigma_\epsilon G_\epsilon^{-1}) + \tfrac{1}{2}\mathrm{grad}\, H_\epsilon : \Sigma_\epsilon + O(\epsilon).
\end{aligned}$$

The projections into the normal bundle and its orthogonal complement can be estimated by

$$\epsilon^{-2}Q_\epsilon F_U(x_\epsilon) = O(1), \quad \text{resp.} \quad \epsilon^{-2}P_\epsilon F_U(x_\epsilon) = \epsilon^{-2}H_\epsilon z_\epsilon + O(1).$$

Proof. We use the fact that $z_\epsilon = O(\epsilon)$ and a long computation:

$$\begin{aligned}
F^i_U(x_\epsilon) &= g^{ij}(x_\epsilon)\frac{\partial U(x_\epsilon)}{\partial x^j} = g^{ij}(x_\epsilon)\left(\frac{\partial U(x_\epsilon)}{\partial x^j} - \frac{\partial U(y_\epsilon)}{\partial x^j}\right) \\
&= g^{ij}(x_\epsilon)\left(\frac{\partial^2 U(y_\epsilon)}{\partial x^j\partial x^k}z^k_\epsilon + \frac{1}{2}\frac{\partial^3 U(y_\epsilon)}{\partial x^j\partial x^k\partial x^l}z^k_\epsilon z^l_\epsilon\right) + O(\epsilon^3) \\
&= s_1 + s_2 + s_3 + O(\epsilon^3),
\end{aligned}$$

where we have split into the following three terms:

$$s_1 = g^{ij}(y_\epsilon)\frac{\partial^2 U(y_\epsilon)}{\partial x^j\partial x^k}z^k_\epsilon = H^i_k(y_\epsilon)z^k_\epsilon, \quad s_2 = (g^{ij}(x_\epsilon) - g^{ij}(y_\epsilon))\frac{\partial^2 U(y_\epsilon)}{\partial x^j\partial x^k}z^k_\epsilon,$$

and

$$s_3 = \tfrac{1}{2}g^{ij}(x_\epsilon)\frac{\partial^3 U(y_\epsilon)}{\partial x^j\partial x^k\partial x^l}z^k_\epsilon z^l_\epsilon = \tfrac{1}{2}g^{ij}(y_\epsilon)\frac{\partial^3 U(y_\epsilon)}{\partial x^j\partial x^k\partial x^l}z^k_\epsilon z^l_\epsilon + O(\epsilon^3).$$

The last two need further treatment. By differentiating the relation $I = GG^{-1}$, we get

$$DG^{-1} = -G^{-1}(DG)G^{-1}, \quad \text{i.e.,} \quad \frac{\partial g^{ij}}{\partial x^l} = -g^{iq}\frac{\partial g_{qp}}{\partial x^l}g^{pj}.$$

Thus, a Taylor expansion of $g^{ij}(x_\epsilon) - g^{ij}(y_\epsilon)$ in the expression of s_2 yields

$$s_2 = -g^{iq}(y_\epsilon)\frac{\partial g_{qp}(y_\epsilon)}{\partial x^l}g^{pj}(y_\epsilon)\frac{\partial^2 U(y_\epsilon)}{\partial x^j \partial x^k}z_\epsilon^k z_\epsilon^l + O(\epsilon^3)$$

$$= -g^{iq}(y_\epsilon)\frac{\partial g_{qp}(y_\epsilon)}{\partial x^l}H_k^p(y_\epsilon)z_\epsilon^k z_\epsilon^l + O(\epsilon^3).$$

Likewise, by differentiating

$$\frac{\partial^2 U}{\partial x^k \partial x^l} = g_{kq}H_l^q,$$

we obtain

$$\frac{\partial^3 U}{\partial x^j \partial x^k \partial x^l} = \frac{\partial g_{kq}}{\partial x^j}H_l^q + g_{kq}\frac{\partial H_l^q}{\partial x^j}.$$

Inserting that into the expression for s_3 yields

$$s_3 = \tfrac{1}{2}g^{ij}(y_\epsilon)\frac{\partial g_{kq}(y_\epsilon)}{\partial x^j}H_l^q(y_\epsilon)z_\epsilon^k z_\epsilon^l + \tfrac{1}{2}g^{ij}(y_\epsilon)g_{kq}(y_\epsilon)\frac{\partial H_l^q(y_\epsilon)}{\partial x^j}z_\epsilon^k z_\epsilon^l + O(\epsilon^3).$$

Altogether, after regrouping and renaming some summation indices we get

$$F_U^i(x_\epsilon) =$$

$$H_j^i(y_\epsilon)z_\epsilon^j - \hat{\Gamma}_{jk}^i(y_\epsilon)H_l^j(y_\epsilon)z_\epsilon^k z_\epsilon^l + \tfrac{1}{2}g^{ij}(y_\epsilon)\frac{\partial H_k^l(y_\epsilon)}{\partial x^j}z_\epsilon^k z_\epsilon^q g_{ql}(y_\epsilon) + O(\epsilon^3),$$

which immediately gives the desired formula. The assertions on the projections follow from the nondegeneracy conditions (II.7).   □

§2.2.   Step 2: The Weak Virial Theorem

This step of the proof is about the distribution of energy in the constrained and the normal motion.

Definition 5. The kinetic energy $T_\epsilon^\|$, the potential energy $U_\epsilon^\|$, and the total energy $E_\epsilon^\|$ of the constrained motion are defined by

$$T_\epsilon^\| = \tfrac{1}{2}g_{ij}(y_\epsilon)\dot{y}_\epsilon^i \dot{y}_\epsilon^j, \qquad U_\epsilon^\| = V(y_\epsilon), \qquad E_\epsilon^\| = T_\epsilon^\| + U_\epsilon^\|.$$

The kinetic energy $T_\epsilon^\perp$, the potential energy $U_\epsilon^\perp$, and the total energy $E_\epsilon^\perp$ of the normal motion are defined by

$$T_\epsilon^\perp = \tfrac{1}{2}g_{ij}(y_\epsilon)\dot{z}_\epsilon^i \dot{z}_\epsilon^j, \qquad U_\epsilon^\perp = \tfrac{1}{2}\epsilon^{-2}\langle H(y_\epsilon)z_\epsilon, z_\epsilon\rangle, \qquad E_\epsilon^\perp = T_\epsilon^\perp + U_\epsilon^\perp.$$

Notice that only the potential parts are defined in a coordinate independent fashion. This ambiguity, however, will disappear in the limit $\epsilon \to 0$ as an implication of the weak virial theorem.

The quadratic expressions Π_ϵ and Σ_ϵ allow elegant expressions for the energies of the normal motion and their weak* limits in $L^\infty[0,T]$,

$$T_\epsilon^\perp = \tfrac{1}{2}\operatorname{tr}\Pi_\epsilon \overset{*}{\rightharpoonup} T_0^\perp = \tfrac{1}{2}\operatorname{tr}\Pi_0,$$
$$U_\epsilon^\perp = \tfrac{1}{2}\operatorname{tr}(H_\epsilon\Sigma_\epsilon) \overset{*}{\rightharpoonup} U_0^\perp = \tfrac{1}{2}\operatorname{tr}(H_0\Sigma_0). \tag{II.26}$$

Recall that we use the notation $H_\epsilon = H(y_\epsilon)$.

The following lemma shows that we essentially have split all the energy into the constrained and normal motion.

Lemma 6. *The total energy E_ϵ decomposes into*

$$E_\epsilon = E_\epsilon^\| + E_\epsilon^\perp + O(\epsilon).$$

Both parts converge uniformly as functions in $C[0,T]$,

$$E_\epsilon^\| \to E_0^\| = \tfrac{1}{2}\langle \dot{x}_0, \dot{x}_0\rangle + V(x_0), \qquad E_\epsilon^\perp \to E_0^\perp = E_0 - E_0^\|.$$

Proof. Differentiation of the orthogonality relation

$$\langle \dot{y}_\epsilon, z_\epsilon \rangle = g_{ij}(y_\epsilon)\dot{y}_\epsilon^i z_\epsilon^j = 0$$

with respect to time yields

$$g_{ij}(y_\epsilon)\ddot{y}_\epsilon^i \dot{z}_\epsilon^j = -g_{ij}(y_\epsilon)\ddot{y}_\epsilon^i z_\epsilon^j - \frac{\partial g_{ij}(y_\epsilon)}{\partial x^k}\dot{y}_\epsilon^i \dot{y}_\epsilon^k z_\epsilon^j = O(\epsilon).$$

Thus, the kinetic energy decomposes into

$$\tfrac{1}{2}\langle \dot{x}_\epsilon, \dot{x}_\epsilon \rangle = \tfrac{1}{2}g_{ij}(x_\epsilon)\dot{x}_\epsilon^i \dot{x}_\epsilon^j = T_\epsilon^\| + T_\epsilon^\perp + O(\epsilon).$$

Taylor expansion of the potentials gives

$$V(x_\epsilon) = U_\epsilon^\| + O(\epsilon)$$

and

$$\epsilon^{-2}U(x_\epsilon) = \tfrac{1}{2}\epsilon^{-2}\frac{\partial^2 U(y_\epsilon)}{\partial x^i \partial x^k}z_\epsilon^i z_\epsilon^j + O(\epsilon) = U_\epsilon^\perp + O(\epsilon).$$

Hence, the energy decomposes as claimed.

Lemma 3 implies the asserted uniform convergence of the constrained part $E_\epsilon^\|$ of the energy. By Eq. (II.3), the total energy converges as a number in $\mathbb{R}$, $E_\epsilon \to E_0$. This readily implies the uniform convergence of the normal part $E_\epsilon^\perp$ of the energy.   □

Now, we state and prove a central result of our argument. For reasons we have discussed at length in §I.2.6, we call it the *weak virial theorem*.

Lemma 7. (Weak Virial Theorem). *The limit of the quadratic expressions of normal position and velocity are related by the Hessian,*

$$\Pi_0 = H_0 \Sigma_0. \tag{II.27}$$

In the limit, the quadratic expressions commute with the Hessian,

$$[H_0, \Sigma_0] = 0, \qquad [H_0, \Pi_0] = 0, \tag{II.28}$$

and there is an equi-partitioning of the kinetic and potential energy of the normal motion,

$$T_0^{\perp} = U_0^{\perp}.$$

Proof. [28] By Lemma 2, we obtain

$$\Xi_\epsilon = \dot{z}_\epsilon \otimes z_\epsilon G_\epsilon = O(\epsilon) \;\; \rightarrow \; 0,$$

and that this quantity Ξ_ϵ has a *bounded* time derivative,

$$\dot{\Xi}_\epsilon = \dot{z}_\epsilon \otimes \dot{z}_\epsilon G_\epsilon + \ddot{z}_\epsilon \otimes z_\epsilon G_\epsilon + \dot{z}_\epsilon \otimes z_\epsilon \dot{G}_\epsilon = \Pi_\epsilon - H_\epsilon \Sigma_\epsilon + O(\epsilon). \tag{II.29}$$

In the last step, we have made use of the second order equation (II.25) of Lemma 3. By Principle I.1, taking weak* limits on both sides of Eq. (II.29) yields

$$0 = \Pi_0 - H_0 \Sigma_0.$$

This readily implies that, in the limit, the kinetic and potential parts of the normal energy are equal,

$$T_0^{\perp} = \tfrac{1}{2} \operatorname{tr}(\Pi_0) = \tfrac{1}{2} \operatorname{tr}(H_0 \Sigma_0) = U_0^{\perp}.$$

Since all the tensors Π_0, Σ_0, and H_0 are linear operators $T_{x_0} M \rightarrow T_{x_0} M$ and they are obviously selfadjoint with respect to the Riemannian metric, we get the commutation relations (II.28). $\qquad \Box$

From the relation (II.27) of the weak virial theorem, an important fact follows: In contrast to the coordinate-dependent definition of Π_ϵ, the limit Π_0 represents—as its counterparts H_0 and Σ_0 do—an *invariantly* defined field of linear maps $T_{x_0} M \rightarrow T_{x_0} M$. In the same way, the limit normal part $T_0^{\perp}$ of the kinetic energy is a scalar field invariantly defined along x_0.

The weak virial theorem allows us to give a partial, abstract, and general answer to the homogenization problem.

---

[28]This proof owes its *formal structure* to the traditional proof of the virial theorem of classical mechanics which may be found for instance in the textbooks of ABRAHAM and MARSDEN [1, Theorem 3.7.30], GOLDSTEIN [39, Chap. 3-4], and LANDAU and LIFSHITZ [62, p. 23].

A *different* proof, using the localization principle of *semiclassical measures*, can be found in Appendix D. That considerably more technical, though still short proof systematizes the result and broadens the perspective by relating it, for instance, to the theory of compensated compactness.

Lemma 8. (The Abstract Limit Equation). *The limit x_0 obeys*

$$\ddot{x}_0 + \Gamma(x_0)(\dot{x}_0, \dot{x}_0) + F_V(x_0) + F_U^{\text{hom}}(t) \perp T_{x_0}N, \qquad \text{(II.30)}$$

which is the second order equation of motion for a mechanical system that is constrained to the manifold N. The homogenized force $F_U^{\text{hom}}(t)$ is given by

$$F_U^{\text{hom}}(t) = \tfrac{1}{2}\operatorname{grad} H(x_0(t)) : \Sigma_0(t). \qquad \text{(II.31)}$$

Proof. As in the proof of Lemma 3, we multiply the equation of motion (II.20) by the projection Q_ϵ. However, this time we are interested in the weak* limit behavior of each term and not only in their boundedness. We get

$$Q_\epsilon \ddot{y}_\epsilon + Q_\epsilon \ddot{z}_\epsilon + Q_\epsilon \Gamma(x_\epsilon)(\dot{y}_\epsilon, \dot{y}_\epsilon) + 2 Q_\epsilon \Gamma(x_\epsilon)(\dot{y}_\epsilon, \dot{z}_\epsilon) + Q_\epsilon \hat{\Gamma}(x_\epsilon)(\dot{z}_\epsilon, \dot{z}_\epsilon)$$
$$+ Q_\epsilon F_V(x_\epsilon) + \epsilon^{-2} Q_\epsilon F_U(x_\epsilon) = 0. \qquad \text{(II.32)}$$

Because of the weak* convergence $\dot{z}_\epsilon \overset{*}{\rightharpoonup} 0$ and the uniform convergences $x_\epsilon \to x_0$ and $\dot{y}_\epsilon \to \dot{x}_0$, we get

$$Q_\epsilon \Gamma(x_\epsilon)(\dot{y}_\epsilon, \dot{z}_\epsilon) \overset{*}{\rightharpoonup} 0.$$

Using Eq. (II.23) and Eq. (II.24), we obtain

$$Q_\epsilon \ddot{z}_\epsilon = 2 Q_\epsilon \dot{P}_\epsilon \dot{z}_\epsilon + Q_\epsilon \dot{P}_\epsilon Q_\epsilon \dot{x}_\epsilon + O(\epsilon) \overset{*}{\rightharpoonup} 0$$

since $Q_\epsilon \dot{P}_\epsilon Q_\epsilon = 0$ by Lemma 9 below. We get further

$$Q_\epsilon \hat{\Gamma}(x_\epsilon)(\dot{z}_\epsilon, \dot{z}_\epsilon) = Q_\epsilon \hat{\Gamma}(x_\epsilon) : (\Pi_\epsilon G_\epsilon^{-1}) \overset{*}{\rightharpoonup} Q_0 \hat{\Gamma}_0 : (\Pi_0 G_0^{-1}).$$

Lemma 5 and the nondegeneracy condition (II.7) give

$$\epsilon^{-2} Q_\epsilon F_U(x_\epsilon) = -Q_\epsilon \hat{\Gamma}_\epsilon : (H_\epsilon \Sigma_\epsilon G_\epsilon^{-1}) + \tfrac{1}{2}\operatorname{grad} H_\epsilon : \Sigma_\epsilon + O(\epsilon)$$
$$\overset{*}{\rightharpoonup} -Q_0 \hat{\Gamma}_0 : (H_0 \Sigma_0 G_0^{-1}) + \tfrac{1}{2}\operatorname{grad} H_0 : \Sigma_0.$$

Thus, taking weak* limits in Eq. (II.32) yields the equation

$$Q_0 \left(\ddot{x}_0 + \Gamma(x_0)(\dot{x}_0, \dot{x}_0) + F_V(x_0) + \tfrac{1}{2}\operatorname{grad} H(x_0) : \Sigma_0 \right)$$
$$+ Q_0 \hat{\Gamma}_0 : ((\Pi_0 - H_0 \Sigma_0)G_0^{-1}) = 0,$$

where the last term is *zero*[29] because of the weak virial theorem (II.27). $\square$

We call the limit equation (II.30) *abstract* since we have no real access to the time-dependent force field F_U^{hom}. It involves the weak* limit Σ^0, an expression that depends on the particular subsequence we have chosen. Therefore, we *cannot* even conclude that x_0 is *unique*, i.e., independent of the chosen subsequence of ϵ.

---

[29]This term would be trivially zero for a flat manifold M, like the Euclidean space, since then $\hat{\Gamma} \equiv 0$. Here is the first of two places in our proof, where such a metric-dependent term drops out like magic because of the weak virial theorem.

Remark 1. For $M = \mathbb{R}^m$ endowed with the Euclidean metric, Lemma 8 has been proven by BORNEMANN and SCHÜTTE [18, Theorem 2.1]. For critical submanifolds N with codimension $r = 1$ this lemma appears in somewhat different form—using suitable *averaging operators*—in the work of the physicists KOPPE and JENSEN [58, Eq. (5)] and VAN KAMPEN [49, Eq. (8.33)].

For the purposes of Chapter IV, we state the following lemma more general than we have needed in the proof of Lemma 8 above.

Lemma 9. *Let P be a projection operator in some Hilbert space $\mathscr{H}$, and Q be an arbitrary bounded operator, both depending continuously differentiable on a parameter. Assume that*

$$PQ = QP = 0.$$

Then, if we denote differentiation with respect to the parameter by a prime, we obtain

$$PQ'P = 0, \qquad PP'P = 0.$$

Proof. Differentiation of $PQ = 0$ gives $P'Q + PQ' = 0$. Because of $QP = 0$, multiplication by P from the right yields $PQ'P = 0$. Likewise, by differentiating the projection relation $P^2 = P$, we get $P'P + PP' = P'$. Multiplication by P from the right gives $PP'P = 0$. □

§2.3. Step 3: Adiabatic Invariance of the Normal Actions

Until now we have not made any use of the smooth spectral decomposition of the Hessian H introduced in §1.3. Recall that the corresponding eigen-projections P_λ induce an orthogonal decomposition of the normal spaces $T_y N^\perp$, $y \in N$. In particular, we decompose the normal motion into[30]

$$z_\epsilon = \sum_\lambda z_{\epsilon\lambda} \qquad \text{with} \qquad z_{\epsilon\lambda} = P_{\epsilon\lambda} z_\epsilon.$$

As we have done with the projection P in §2.1, we may view the projections P_λ in the tubular coordinate system as parameter-dependent matrices. This allows to extend their action to all vectors of $\mathbb{R}^m$ in a coordinate-dependent fashion. Analogously to the relations (II.23) and (II.24), we now get

$$\dot{z}_{\epsilon\lambda} = P_{\epsilon\lambda}\dot{z}_\epsilon + \dot{P}_{\epsilon\lambda} z_\epsilon = P_{\epsilon\lambda}\dot{z}_\epsilon + O(\epsilon) \tag{II.33}$$

and

$$\ddot{z}_{\epsilon\lambda} = P_{\epsilon\lambda}\ddot{z}_\epsilon + 2\dot{P}_{\epsilon\lambda}\dot{z}_\epsilon + O(\epsilon) = P_{\epsilon\lambda}\ddot{z}_\epsilon + O(1). \tag{II.34}$$

---

[30] As usual, we abbreviate $P_{\epsilon\lambda} = P_\lambda(y_\epsilon)$ and $\omega_{\epsilon\lambda} = \omega_\lambda(y_\epsilon)$, including $\epsilon = 0$.

Therefore, multiplying the differential equation (II.25) by $P_{\epsilon\lambda}$ immediately yields the component-wise relation

$$\ddot{z}_{\epsilon\lambda} + \epsilon^{-2}\omega_{\epsilon\lambda}^2 z_{\epsilon\lambda} = O(1). \qquad (\text{II.35})$$

This equation explains why we have called ω_λ the normal *frequencies* of the Hessian H. The components $z_{\epsilon\lambda}$ are thus *perturbed harmonic oscillations*.

The next lemma shows that the limit quadratic expressions Π_0 and Σ_0 have a block-diagonal form with respect to the eigenspaces of H. This is a direct consequence of the weak virial theorem and a certain resonance condition.

Lemma 10. *Let there be essentially no resonances of order two along the limit x_0, i.e.,*

$$\omega_{0\lambda} \neq \omega_{0\mu} \qquad a.e.$$

for $\lambda \neq \mu$. Then, one gets the block-diagonalizations

$$\Sigma_0 = \sum_\lambda P_{0\lambda}\Sigma_0 P_{0\lambda}, \qquad P_{0\lambda}\Sigma_0 P_{0\mu} = 0 \quad \lambda \neq \mu, \qquad (\text{II.36})$$

and

$$\Pi_0 = \sum_\lambda \omega_{0\lambda}^2 P_{0\lambda}\Sigma_0 P_{0\lambda}, \qquad P_{0\lambda}\Pi_0 P_{0\mu} = 0 \quad \lambda \neq \mu, \qquad (\text{II.37})$$

as functions in $L^\infty([0,T], \mathbb{R}^{m\times m})$.

Proof. Multiplying the commutativity relation $H_0\Sigma_0 = \Sigma_0 H_0$ of the weak virial theorem by the projection $P_{0\lambda}$ from the left and by $P_{0\mu}$ from the right gives

$$\omega_{0\lambda}^2 P_{0\lambda}\Sigma_0 P_{0\mu} = \omega_{0\mu}^2 P_{0\lambda}\Sigma_0 P_{0\mu}.$$

Thus, by the resonance assumption, for $\lambda \neq \mu$ we get $P_{0\lambda}\Sigma_0 P_{0\mu} = 0$ in $L^\infty([0,T], \mathbb{R}^{m\times m})$. The same argument applies to Π_0. The representation $\Pi_0 = H_0\Sigma_0$ of the weak virial theorem yields

$$P_{0\lambda}\Pi_0 = P_{0\lambda}H_0\Sigma_0 = \omega_{0\lambda}^2 P_{0\lambda}\Sigma_0.$$

By $Q_\epsilon z_\epsilon = 0$, we obtain $Q_\epsilon\Sigma_\epsilon = \Sigma_\epsilon Q_\epsilon = 0$ and, taking weak* limits, $Q_0\Sigma_0 = \Sigma_0 Q_0 = 0$. This implies, by using the weak virial theorem, i.e., $\Pi_0 = H_0\Sigma_0 = \Sigma_0 H_0$, that there is $Q_0\Pi_0 = \Pi_0 Q_0 = 0$. Thus, Σ_0 and Π_0 have the asserted block-structure.   □

The corresponding off-diagonal blocks of the non-limit expressions are given by

$$P_{\epsilon\lambda}\Sigma_\epsilon P_{\epsilon\mu} = \epsilon^{-2} z_{\epsilon\lambda} \otimes z_{\epsilon\mu}G_\epsilon, \qquad P_{\epsilon\lambda}\Pi_\epsilon P_{\epsilon\mu} = \dot{z}_{\epsilon\lambda} \otimes \dot{z}_{\epsilon\mu}G_\epsilon + O(\epsilon). \qquad (\text{II.38})$$

These relations can easily be proven in coordinates by using Eq. (II.33) and the selfadjointness $P_\lambda^* = P_\lambda$, i.e.,

$$P_{\lambda i}{}^j g_{jk} = g_{ij} P_{\lambda k}{}^j. \tag{II.39}$$

Thus, Lemma 10 shows the weak* continuity of certain, but certainly not all, quadratic expressions,

$$\epsilon^{-2} z_{\epsilon\lambda} \otimes z_{\epsilon\mu} \overset{*}{\rightharpoonup} 0, \qquad \dot{z}_{\epsilon\lambda} \otimes \dot{z}_{\epsilon\mu} \overset{*}{\rightharpoonup} 0, \qquad \lambda \neq \mu. \tag{II.40}$$

Recall that $\epsilon^{-1} z_\epsilon \overset{*}{\rightharpoonup} 0$ and $\dot{z}_\epsilon \overset{*}{\rightharpoonup} 0$.

Surprisingly, only very little is needed to establish full *cubic weak continuity*. A further resonance condition and the perturbed harmonic oscillator equation (II.35) turn out to be enough.

Lemma 11. *Let there be essentially no resonances of order three along the limit x_0, i.e.,*

$$\omega_{0\lambda} + \omega_{0\mu} \neq \omega_{0\nu} \qquad a.e.$$

for all indices λ, μ, and ν. Then, there is weak* convergence of cubic terms,

$$\epsilon^{-2} z_\epsilon \otimes z_\epsilon \otimes \dot{z}_\epsilon \overset{*}{\rightharpoonup} 0, \qquad \dot{z}_\epsilon \otimes \dot{z}_\epsilon \otimes \dot{z}_\epsilon \overset{*}{\rightharpoonup} 0, \tag{II.41}$$

in components,

$$\epsilon^{-2} z_{\epsilon\lambda}^i z_{\epsilon\mu}^j \dot{z}_{\epsilon\nu}^k \overset{*}{\rightharpoonup} 0, \qquad \dot{z}_{\epsilon\lambda}^i \dot{z}_{\epsilon\mu}^j \dot{z}_{\epsilon\nu}^k \overset{*}{\rightharpoonup} 0, \tag{II.42}$$

for all indices λ, μ, and ν, and i, j, and k.

Proof. Differentiating the uniform limit

$$z_{\epsilon\lambda}^i \dot{z}_{\epsilon\mu}^j \dot{z}_{\epsilon\nu}^k \to 0$$

gives, by using Eq. (II.35) and Principle I.1,

$$\dot{z}_{\epsilon\lambda}^i \dot{z}_{\epsilon\mu}^j \dot{z}_{\epsilon\nu}^k - \epsilon^{-2} \omega_{\epsilon\mu}^2 z_{\epsilon\lambda}^i z_{\epsilon\mu}^j \dot{z}_{\epsilon\nu}^k - \epsilon^{-2} \omega_{\epsilon\nu}^2 z_{\epsilon\lambda}^i \dot{z}_{\epsilon\mu}^j z_{\epsilon\nu}^k \overset{*}{\rightharpoonup} 0. \tag{II.43}$$

On the other hand, differentiating the uniform limit

$$\epsilon^{-2} z_{\epsilon\lambda}^i z_{\epsilon\mu}^j z_{\epsilon\nu}^k \to 0,$$

and applying Principle I.1 once more, immediately yields

$$\epsilon^{-2} (\dot{z}_{\epsilon\lambda}^i z_{\epsilon\mu}^j z_{\epsilon\nu}^k + z_{\epsilon\lambda}^i \dot{z}_{\epsilon\mu}^j z_{\epsilon\nu}^k + z_{\epsilon\lambda}^i z_{\epsilon\mu}^j \dot{z}_{\epsilon\nu}^k) \overset{*}{\rightharpoonup} 0. \tag{II.44}$$

By combining the limit equation (II.44) with the three limit equations obtained from Eq. (II.43) by a cyclic permutation of the indices i, j, and k, and simultaneously of the indices λ, μ, and ν, we get

$$\underbrace{\begin{pmatrix} 0 & 1 & 1 & 1 \\ 1 & 0 & -\omega_{\epsilon\nu}^2 & -\omega_{\epsilon\mu}^2 \\ 1 & -\omega_{\epsilon\nu}^2 & 0 & -\omega_{\epsilon\lambda}^2 \\ 1 & -\omega_{\epsilon\mu}^2 & -\omega_{\epsilon\lambda}^2 & 0 \end{pmatrix}}_{=A_\epsilon} \begin{pmatrix} \dot{z}_{\epsilon\lambda}^i \dot{z}_{\epsilon\mu}^j \dot{z}_{\epsilon\nu}^k \\ \epsilon^{-2} \dot{z}_{\epsilon\lambda}^i z_{\epsilon\mu}^j z_{\epsilon\nu}^k \\ \epsilon^{-2} z_{\epsilon\lambda}^i \dot{z}_{\epsilon\mu}^j z_{\epsilon\nu}^k \\ \epsilon^{-2} z_{\epsilon\lambda}^i z_{\epsilon\mu}^j \dot{z}_{\epsilon\nu}^k \end{pmatrix} \overset{*}{\rightharpoonup} 0.$$

Now, we have that

$$\det A_\epsilon = (\omega_{\epsilon\lambda} + \omega_{\epsilon\mu} + \omega_{\epsilon\nu})(\omega_{\epsilon\lambda} - \omega_{\epsilon\mu} - \omega_{\epsilon\nu})(\omega_{\epsilon\mu} - \omega_{\epsilon\nu} - \omega_{\epsilon\lambda})(\omega_{\epsilon\nu} - \omega_{\epsilon\lambda} - \omega_{\epsilon\mu}).$$

Therefore, the resonance condition yields $\det A_0 \neq 0$, almost everywhere, which finishes the proof.   □

Remark 2. There is a nice motivation for the off-diagonal quadratic weak* continuity (II.40) and the cubic weak* continuity (II.41). Suppose, the components $z_{\epsilon\lambda}$ satisfy *unperturbed* harmonic oscillator equations,

$$\ddot{z}_{\epsilon\lambda} + \epsilon^{-2}\omega_\lambda^2 z_{\epsilon\lambda} = 0$$

with *constant* frequencies ω_λ. Using the formulas for products of trigonometric functions, one writes the terms $\dot{z}_{\lambda\epsilon} \otimes \dot{z}_{\mu\epsilon}$ as a linear combination of the harmonic oscillations

$$\exp\left(i(\omega_\lambda - \omega_\mu)t/\epsilon\right) \quad \text{and} \quad \exp\left(i(\omega_\lambda + \omega_\mu)t/\epsilon\right).$$

If there are no resonances of order two, the Riemann-Lebesgue lemma, [84, Sec. 5.14], shows the weak* convergence to zero for $\lambda \neq \mu$. Correspondingly, a further application of the product formulas yields a representation of the terms $\dot{z}_{\lambda\epsilon} \otimes \dot{z}_{\mu\epsilon} \otimes \dot{z}_{\epsilon\nu}$ as a linear combinations of the harmonic oscillations

$$\exp\left(\pm i(\omega_\lambda + \omega_\mu - \omega_\nu)t/\epsilon\right)$$

over all cyclic permutations of λ, μ, and ν. Thus, if there are no resonances of order three, a further application of the Riemann-Lebesgue lemma shows the weak* convergence to zero. Basically, this is the argument given by TAKENS in his paper [94]. However, one has to modify the idea to incorporate the perturbations which are present in Eq. (II.35). This is only sketched in [94]; working out the details appears to be quite involved. We think that our proof of the weak* continuity of cubic expressions has the advantage of being much simpler—besides being more general with respect to the resonance conditions.

Later, we will need a more *detailed* knowledge of the $O(1)$-term on the right-hand-side of Eq. (II.35).

Lemma 12. *The perturbed harmonic oscillator equation for the component $z_{\epsilon\lambda}$ is given by*

$$
\ddot{z}_{\epsilon\lambda}^i + \epsilon^{-2}\omega_{\epsilon\lambda}^2 z_{\epsilon\lambda}^i = 2(\dot{P}_{\epsilon\lambda}\dot{z}_\epsilon)^i - 2(P_{\epsilon\lambda}\Gamma_\epsilon(\dot{y}_\epsilon, \dot{z}_\epsilon))^i \\
+ a_{\epsilon\lambda jk}{}^i \dot{z}_\epsilon^j \dot{z}_\epsilon^k + \epsilon^{-2}b_{\epsilon\lambda jk}{}^i z_\epsilon^j z_\epsilon^k + c_{\epsilon\lambda}{}^i
\tag{II.45}
$$

with uniformly converging fields $a_{\epsilon\lambda}$, $b_{\epsilon\lambda}$, and $c_{\epsilon\lambda}$.

Proof. We multiply the equation of motion (II.20) by the projection $P_{\epsilon\lambda}$ from the left obtaining

$$P_{\epsilon\lambda}\ddot{x}_\epsilon + P_{\epsilon\lambda}\Gamma(x_\epsilon)(\dot{x}_\epsilon, \dot{x}_\epsilon) + P_{\epsilon\lambda}F_V(x_\epsilon) + \epsilon^{-2}P_{\epsilon\lambda}F_U(x_\epsilon) = 0.$$

Each term of this equation will be studied separately. In order to simplify, we use the notation

$$g(t; \epsilon) \equiv f(t; \epsilon) \quad (\text{mod } C^0\text{-lim})$$

to mean that the difference $g(t; \epsilon) - f(t; \epsilon)$ converges uniformly in C^0. We get by $P_{\epsilon\lambda}P_\epsilon = P_{\epsilon\lambda}$ and the relations (II.23), (II.24) as well as (II.33), (II.34) that

$$
\begin{aligned}
P_{\epsilon\lambda}\ddot{x}_\epsilon &= P_{\epsilon\lambda}P_\epsilon\ddot{x}_\epsilon = P_{\epsilon\lambda}\ddot{z}_\epsilon - 2P_{\epsilon\lambda}\dot{P}_\epsilon\dot{z}_\epsilon - P_{\epsilon\lambda}\dot{P}_\epsilon\dot{y}_\epsilon + O(\epsilon) \\
&= \ddot{z}_{\epsilon\lambda} - 2\underbrace{\left(\dot{P}_{\epsilon\lambda}P_\epsilon + P_{\epsilon\lambda}\dot{P}_\epsilon\right)}_{=\dot{P}_{\epsilon\lambda}}\dot{z}_\epsilon - P_{\epsilon\lambda}\dot{P}_\epsilon\dot{y}_\epsilon + O(\epsilon) \\
&\equiv \ddot{z}_{\epsilon\lambda} - 2\dot{P}_{\epsilon\lambda}\dot{z}_\epsilon \quad (\text{mod } C^0\text{-lim}).
\end{aligned}
$$

The symmetric bilinearity of Γ yields

$$P_{\epsilon\lambda}\Gamma(x_\epsilon)(\dot{x}_\epsilon, \dot{x}_\epsilon) \equiv 2P_{\epsilon\lambda}\Gamma_\epsilon(\dot{y}_\epsilon, \dot{z}_\epsilon) + P_{\epsilon\lambda}\Gamma_\epsilon(\dot{z}_\epsilon, \dot{z}_\epsilon) \quad (\text{mod } C^0\text{-lim}).$$

Obviously, we obtain

$$P_{\epsilon\lambda}F_V(x_\epsilon) \equiv 0 \quad (\text{mod } C^0\text{-lim}).$$

Finally, Lemma 5 and the spectral decomposition (II.8) gives

$$
\begin{aligned}
\epsilon^{-2}P_{\epsilon\lambda}F_U(x_\epsilon) &\equiv \epsilon^{-2}\omega_{\epsilon\lambda}z_{\epsilon\lambda} - P_{\epsilon\lambda}\hat{\Gamma}_\epsilon : (H_\epsilon\Sigma_\epsilon G_\epsilon^{-1}) \\
&\quad + \tfrac{1}{2}P_{\epsilon\lambda}\operatorname{grad} H_\epsilon : \Sigma_\epsilon \quad (\text{mod } C^0\text{-lim}).
\end{aligned}
$$

Putting the results together, we obtain the asserted equation.   □

Now, we split the energy of the normal motion into its components, corresponding to the splitting $z_\epsilon = z_{\epsilon 1} + \ldots + z_{\epsilon s}$.

Definition 6. The kinetic energy $T_{\epsilon\lambda}^\perp$, the potential energy $U_{\epsilon\lambda}^\perp$, and the total energy $E_{\epsilon\lambda}^\perp$ of the normal λ-component are defined as

$$T_{\epsilon\lambda}^\perp = \tfrac{1}{2}g_{ij}(y_\epsilon)\dot{z}_{\epsilon\lambda}^i\dot{z}_{\epsilon\lambda}^j, \quad U_{\epsilon\lambda}^\perp = \tfrac{1}{2}\epsilon^{-2}\omega_{\epsilon\lambda}^2\langle z_{\epsilon\lambda}, z_{\epsilon\lambda}\rangle, \quad E_{\epsilon\lambda}^\perp = T_{\epsilon\lambda}^\perp + U_{\epsilon\lambda}^\perp.$$

Again, only the potential parts are invariantly defined in a coordinate independent fashion. This ambiguity will disappear in the limit $\epsilon \to 0$, because of the weak virial theorem.

Lemma 13. *The energy of the normal motion splits as*

$$E_\epsilon^\perp = \sum_\lambda E_{\epsilon\lambda}^\perp + O(\epsilon).$$

By defining $\sigma_\lambda = \mathrm{tr}(P_{0\lambda}\Sigma_0 P_{0\lambda})$, one obtains the weak limits*

$$T_{\epsilon\lambda}^\perp \overset{*}{\rightharpoonup} T_{0\lambda}^\perp = \tfrac{1}{2}\omega_{0\lambda}^2\sigma_\lambda, \quad U_{\epsilon\lambda}^\perp \overset{*}{\rightharpoonup} U_{0\lambda}^\perp = \tfrac{1}{2}\omega_{0\lambda}^2\sigma_\lambda, \quad E_{\epsilon\lambda}^\perp \overset{*}{\rightharpoonup} E_{0\lambda}^\perp = \omega_{0\lambda}^2\sigma_\lambda,$$

such that there is an equi-partitioning of energy in the limit.

Proof. By using properties of the trace and the relation (II.38), the kinetic energy splits as follows,

$$\begin{aligned} T_\epsilon^\perp = \tfrac{1}{2}\mathrm{tr}\,\Pi_\epsilon &= \tfrac{1}{2}\sum_{\lambda\mu}\mathrm{tr}(P_{\epsilon\lambda}\Pi_\epsilon P_{\epsilon\mu}) = \tfrac{1}{2}\sum_{\lambda\mu}\mathrm{tr}(\Pi_\epsilon P_{\epsilon\mu}P_{\epsilon\lambda}) \\ &= \tfrac{1}{2}\sum_\lambda\mathrm{tr}(P_{\epsilon\lambda}\Pi_\epsilon P_{\epsilon\lambda}) = \sum_\lambda T_{\epsilon\lambda}^\perp + O(\epsilon). \end{aligned}$$

Likewise, we get

$$U_\epsilon^\perp = \tfrac{1}{2}\mathrm{tr}(H_\epsilon\Sigma_\epsilon) = \tfrac{1}{2}\sum_\lambda \omega_{\epsilon\lambda}^2\,\mathrm{tr}(P_{\epsilon\lambda}\Sigma_\epsilon P_{\epsilon\lambda}) = \sum_\lambda U_{\epsilon\lambda}^\perp.$$

Further, relation (II.38) and $P_{0\lambda}\Pi_0 P_{0\lambda} = \omega_{0\lambda}^2 P_{0\lambda}\Sigma_0 P_{0\lambda}$ which directly follows from the weak virial theorem, Eq. (II.27), yield

$$T_{\epsilon\lambda}^\perp = \tfrac{1}{2}\mathrm{tr}(P_{\epsilon\lambda}\Pi_\epsilon P_{\epsilon\lambda}) + O(\epsilon) \overset{*}{\rightharpoonup} \tfrac{1}{2}\mathrm{tr}(P_{0\lambda}\Pi_0 P_{0\lambda}) = \tfrac{1}{2}\omega_{0\lambda}^2\sigma_\lambda.$$

More directly, we obtain

$$U_{\epsilon\lambda}^\perp = \tfrac{1}{2}\omega_{\epsilon\lambda}^2\,\mathrm{tr}(P_{\epsilon\lambda}\Sigma_\epsilon P_{\epsilon\lambda}) \overset{*}{\rightharpoonup} \tfrac{1}{2}\omega_{0\lambda}^2\sigma_\lambda,$$

which finishes the proof. □

After all these preparations, we come to the main result of the present step in the proof of the homogenization result, Theorem 1. First, we introduce an important notion.

Definition 7. The *action* of the normal λ-component is defined as the time-dependent ratio

$$\theta_\epsilon^\lambda = \frac{E_{\epsilon\lambda}^\perp}{\omega_{\epsilon\lambda}}.$$

Due to the nondegeneracy condition (II.10), this ratio is well-defined.

Lemma 14. *Let there be essentially no resonances of order two or three along the limit x_0. Then, the action of each normal λ-component is an adiabatic invariant,*[31] *i.e.,*

$$\theta_\epsilon^\lambda \;\to\; \theta_*^\lambda = \text{const}$$

uniformly in $C^0[0,T]$. The value of the constant θ_*^λ is given by

$$\theta_*^\lambda = \frac{\langle P_\lambda(x_*)v_*, P_\lambda(x_*)v_* \rangle}{2\,\omega_\lambda(x_*)}.$$

Proof. The main idea in showing the adiabatic invariance is computing the weak* limit of the derivative $\dot{E}_{\epsilon\lambda}^\perp$, where

$$
\begin{aligned}
\dot{E}_{\epsilon\lambda}^\perp \;=\; & \operatorname{tr}\left(\left(\ddot{z}_{\epsilon\lambda} + \epsilon^{-2}\omega_{\epsilon\lambda}^2 z_{\epsilon\lambda}\right) \otimes \dot{z}_{\epsilon\lambda} G_\epsilon\right) \\
& + \tfrac{1}{2}\operatorname{tr}(\dot{z}_{\epsilon\lambda} \otimes \dot{z}_{\epsilon\lambda}\dot{G}_\epsilon) + \tfrac{1}{2}\epsilon^{-2}\omega_{\epsilon\lambda}^2\operatorname{tr}(z_{\epsilon\lambda} \otimes z_{\epsilon\lambda}\dot{G}_\epsilon) \\
& + \tfrac{1}{2}\epsilon^{-2}\operatorname{tr}(z_{\epsilon\lambda} \otimes z_{\epsilon\lambda}G_\epsilon) \cdot \frac{d}{dt}\omega_{\epsilon\lambda}^2.
\end{aligned}
$$

We call the three rows of the right hand side s_1, s_2, and s_3, respectively. The weak* limit of s_3 is given, by using Eq. (II.38) and $\sigma_\lambda = \operatorname{tr}(P_{0\lambda}\Sigma_0 P_{0\lambda})$, as

$$s_3 = \tfrac{1}{2}\operatorname{tr}(P_{\epsilon\lambda}\Sigma_\epsilon P_{\epsilon\lambda}) \cdot \frac{d}{dt}\omega_{\epsilon\lambda}^2 \;\overset{*}{\rightharpoonup}\; \tfrac{1}{2}\sigma_\lambda \frac{d}{dt}\omega_{0\lambda}^2.$$

By using Eqs. (II.38) and (II.8), we obtain

$$
\begin{aligned}
s_2 \;=\; & \tfrac{1}{2}\operatorname{tr}(P_{\epsilon\lambda}\Pi_\epsilon P_{\epsilon\lambda}G_\epsilon^{-1}\dot{G}_\epsilon) + \tfrac{1}{2}\operatorname{tr}(P_{\epsilon\lambda}H_\epsilon\Sigma_\epsilon P_{\epsilon\lambda}G_\epsilon^{-1}\dot{G}_\epsilon) + O(\epsilon) \\
\;\overset{*}{\rightharpoonup}\; & \tfrac{1}{2}\operatorname{tr}(P_{0\lambda}\Pi_0 P_{0\lambda}G_0^{-1}\dot{G}_0) + \tfrac{1}{2}\operatorname{tr}(P_{0\lambda}H_0\Sigma_0 P_{0\lambda}G_0^{-1}\dot{G}_0).
\end{aligned}
$$

The evaluation of the weak* limit of s_1 requires much more work. The detailed second order equation (II.45) of Lemma 12 yields

$$
\begin{aligned}
s_1 \;=\; & \dot{z}_{\epsilon\lambda}^i\left(\ddot{z}_{\epsilon\lambda}^j + \epsilon^{-2}\omega_{\epsilon\lambda}^2 z_{\epsilon\lambda}^j\right)g_{ij}(y_\epsilon) \\
\;=\; & \underbrace{a_{\epsilon\lambda kl}^{\;j}\dot{z}_\epsilon^k \dot{z}_\epsilon^l \dot{z}_{\epsilon\lambda}^i g_{ij}(y_\epsilon)}_{\overset{*}{\rightharpoonup}0} + \underbrace{b_{\epsilon\lambda kl}^{\;j}\epsilon^{-2}z_\epsilon^k z_\epsilon^l \dot{z}_{\epsilon\lambda}^i g_{ij}(y_\epsilon)}_{\overset{*}{\rightharpoonup}0} + \underbrace{c_{\epsilon\lambda}^{\;j}\dot{z}_{\epsilon\lambda}^i g_{ij}(y_\epsilon)}_{\overset{*}{\rightharpoonup}0} \\
& + 2\underbrace{(\dot{P}_{\epsilon\lambda}\dot{z}_\epsilon)^j \dot{z}_{\epsilon\lambda}^i g_{ij}(y_\epsilon)}_{=s_4} - 2\underbrace{(P_{\epsilon\lambda}\Gamma_\epsilon(\dot{y}_\epsilon, \dot{z}_\epsilon))^j \dot{z}_{\epsilon\lambda}^i g_{ij}(y_\epsilon)}_{=s_5}.
\end{aligned}
$$

The first two weak* limits are direct consequences of the cubic weak convergence stated in Lemma 11, the third follows from $\dot{z}_\epsilon \overset{*}{\rightharpoonup} 0$. We recall that

---

[31]Notice that the time interval $[0,T]$ under consideration is of order $O(\epsilon^{-1}\tau_\epsilon)$, where τ_ϵ denotes a typical period of a small oscillation in the normal direction. Thus, the usage of the notion "adiabatic invariant" is in accordance with the definition given in ARNOLD, KOZLOV, and NEISHTADT [6, Chap. 5.4].

$a_{\epsilon\lambda}$, $b_{\epsilon\lambda}$, and $c_{\epsilon\lambda}$ are uniformly convergent. Now, by properties of the trace and the relations (II.33) and (II.38) as well as by the fact that, according to Lemma 9, $P_{\epsilon\lambda}\dot{P}_{\epsilon\lambda}P_{\epsilon\lambda} = 0$, we obtain

$$
\begin{aligned}
s_4 &= \operatorname{tr}(\dot{P}_{\epsilon\lambda}\dot{z}_\epsilon \otimes \dot{z}_{\epsilon\lambda}G_\epsilon) = \sum_\mu \operatorname{tr}(\dot{P}_{\epsilon\lambda}\dot{z}_{\epsilon\mu} \otimes \dot{z}_{\epsilon\lambda}G_\epsilon) \\
&= \sum_\mu \operatorname{tr}(\dot{P}_{\epsilon\lambda}P_{\epsilon\mu}\Pi_\epsilon P_{\epsilon\lambda}) + O(\epsilon) = \sum_{\mu\neq\lambda} \operatorname{tr}(P_{\epsilon\lambda}\dot{P}_{\epsilon\lambda}P_{\epsilon\mu}\Pi_\epsilon) + O(\epsilon) \\
&= \sum_{\mu\neq\lambda} \operatorname{tr}(\dot{P}_{\epsilon\lambda}P_{\epsilon\mu}\Pi_\epsilon P_{\epsilon\lambda}) + O(\epsilon) \overset{*}{\to} \sum_{\mu\neq\lambda} \operatorname{tr}(\dot{P}_{0\lambda}P_{0\mu}\Pi_0 P_{0\lambda}) = 0,
\end{aligned}
$$

where the last equation stems from Eq. (II.37). The next effort involves the quantity s_5. To begin, the selfadjointness relation (II.39) and the projection relation (II.33) yield

$$
\begin{aligned}
s_5 &= 2\Gamma_\epsilon(\dot{y}_\epsilon, \dot{z}_\epsilon)^k P_{\epsilon\lambda k}^{\;j} g_{ji}(y_\epsilon)\dot{z}_{\epsilon\lambda}^i = 2\Gamma_\epsilon(\dot{y}_\epsilon, \dot{z}_\epsilon)^k g_{kj}(y_\epsilon)P_{\epsilon\lambda i}^{\;j}\dot{z}_{\epsilon\lambda}^i \\
&= 2\operatorname{tr}(\Gamma_\epsilon(\dot{y}_\epsilon, \dot{z}_\epsilon) \otimes (P_{\epsilon\lambda}\dot{z}_{\epsilon\lambda})\,G_\epsilon) = 2\operatorname{tr}(\Gamma_\epsilon(\dot{y}_\epsilon, \dot{z}_\epsilon) \otimes \dot{z}_{\epsilon\lambda}\,G_\epsilon) + O(\epsilon).
\end{aligned}
$$

Now, since we can write $\Gamma_\epsilon(\dot{y}_\epsilon, \dot{z}_\epsilon) = A_\epsilon\dot{z}_\epsilon$ with a strongly converging linear operator A_ϵ, we further get by Eq. (II.38)

$$
\begin{aligned}
s_5 &= 2\sum_\mu \operatorname{tr}(A_\epsilon\dot{z}_{\epsilon\mu} \otimes \dot{z}_{\epsilon\lambda}\,G_\epsilon) + O(\epsilon) = 2\sum_\mu \operatorname{tr}(A_\epsilon P_{\epsilon\mu}\Pi_\epsilon P_{\epsilon\lambda}) + O(\epsilon) \\
&= 2\sum_{\mu\neq\lambda} \operatorname{tr}(\underbrace{A_\epsilon P_{\epsilon\mu}\Pi_\epsilon P_{\epsilon\lambda}}_{\overset{*}{\to}A_0 P_{0\mu}\Pi_0 P_{0\lambda}=0}) + \underbrace{2\operatorname{tr}(A_\epsilon P_{\epsilon\lambda}\Pi_\epsilon P_{\epsilon\lambda})}_{=s_6} + O(\epsilon).
\end{aligned}
$$

Finally, we compute the weak* limit of the quantity s_6 as

$$
\begin{aligned}
s_6 &= 2\operatorname{tr}(\Gamma_\epsilon(\dot{y}_\epsilon, \dot{z}_{\epsilon\lambda}) \otimes \dot{z}_{\epsilon\lambda}G_\epsilon) + O(\epsilon) = 2\Gamma_{jk}^i(y_\epsilon)\dot{y}_\epsilon^j \dot{z}_{\epsilon\lambda}^k \dot{z}_{\epsilon\lambda}^l g_{li}(y_\epsilon) + O(\epsilon) \\
&= g_{li}(y_\epsilon)g^{iq}(y_\epsilon)\left(\frac{\partial g_{jq}(y_\epsilon)}{\partial x^k} + \frac{\partial g_{qk}(y_\epsilon)}{\partial x^j} - \frac{\partial g_{jk}(y_\epsilon)}{\partial x^q}\right)\dot{y}_\epsilon^j \dot{z}_{\epsilon\lambda}^k \dot{z}_{\epsilon\lambda}^l + O(\epsilon) \\
&= \left(\underbrace{\frac{\partial g_{jl}(y_\epsilon)}{\partial x^k} - \frac{\partial g_{jk}(y_\epsilon)}{\partial x^l}}_{\text{skewsymmetric in } k \text{ and } l} + \frac{\partial g_{lk}(y_\epsilon)}{\partial x^j}\right)\dot{y}_\epsilon^j \dot{z}_{\epsilon\lambda}^k \dot{z}_{\epsilon\lambda}^l + O(\epsilon) \\
&= \frac{\partial g_{lk}(y_\epsilon)}{\partial x^j}\dot{y}_\epsilon^j \dot{z}_{\epsilon\lambda}^k \dot{z}_{\epsilon\lambda}^l + O(\epsilon) = \operatorname{tr}(\dot{z}_{\epsilon\lambda} \otimes \dot{z}_{\epsilon\lambda}G_\epsilon G_\epsilon^{-1}\dot{G}_\epsilon) + O(\epsilon) \\
&= \operatorname{tr}(P_{\epsilon\lambda}\Pi_\epsilon P_{\epsilon\lambda}G_\epsilon^{-1}\dot{G}_\epsilon) + O(\epsilon) \\
&\overset{*}{\to} \operatorname{tr}(P_{0\lambda}\Pi_0 P_{0\lambda}G_0^{-1}\dot{G}_0).
\end{aligned}
$$

Summarizing, we have shown that

$$s_1 \overset{*}{\rightharpoonup} -\operatorname{tr}(P_{0\lambda}\Pi_0 P_{0\lambda} G_0^{-1}\dot{G}_0).$$

Therefore, we obtain by the weak virial theorem[32]

$$s_1 + s_2 \overset{*}{\rightharpoonup} \tfrac{1}{2}\operatorname{tr}\left(P_{0\lambda}(H_0\Sigma_0 - \Pi_0)P_{0\lambda}G_0^{-1}\dot{G}_0\right) = 0,$$

which implies

$$\dot{E}_{\epsilon\lambda}^{\perp} \overset{*}{\rightharpoonup} \dot{E}_{0\lambda}^{\perp} = \tfrac{1}{2}\sigma_\lambda \frac{d}{dt}\omega_{0\lambda}^2.$$

On the other hand, a direct differentiation of the limit relation given in Lemma 13, i.e., $E_{0\lambda}^{\perp} = \sigma_\lambda \omega_{0\lambda}^2$, yields the expression

$$\dot{E}_{0\lambda}^{\perp} = \dot{\sigma}_\lambda \omega_{0\lambda}^2 + \sigma_\lambda \frac{d}{dt}\omega_{0\lambda}^2.$$

By comparing the two expressions for $\dot{E}_{0\lambda}^{\perp}$, we obtain the following differential equation for the function σ_λ,

$$\frac{\dot{\sigma}_\lambda}{\sigma_\lambda} = -\tfrac{1}{2}\frac{d\omega_{0\lambda}^2/dt}{\omega_{0\lambda}^2} = -\frac{\dot{\omega}_{0\lambda}}{\omega_{0\lambda}}.$$

Solving this differential equation explicitly shows that there is a *constant* θ_*^λ such that

$$\sigma_\lambda = \frac{\theta_*^\lambda}{\omega_\lambda(x_0)}, \qquad E_{0\lambda}^{\perp} = \theta_*^\lambda \omega_\lambda(x_0). \tag{II.46}$$

Now, since the derivative $\dot{E}_{\epsilon\lambda}^{\perp}$ converges weakly, we may conclude—by a further application of the extended Arzelà-Ascoli theorem, Principle I.4— that the energy itself converges *uniformly*. Hence, we get the *uniform* convergence

$$\theta_\epsilon^\lambda = \frac{E_{\epsilon\lambda}^{\perp}}{\omega_{\epsilon\lambda}} \to \theta_0^\lambda = \frac{E_{0\lambda}^{\perp}}{\omega_\lambda(x_0)}, \quad \text{i.e.,} \qquad \theta_0^\lambda \equiv \theta_*^\lambda.$$

In particular, the value of θ_*^λ can be computed for $t = 0$ as the limit of an expression of the initial values. Since $y_\epsilon(0) = x_*$ and $z_\epsilon(0) = 0$ we have

$$U_{\epsilon\lambda}^{\perp}(0) = 0.$$

The projection relations (II.23) and (II.33) together with $P_{\epsilon\lambda}P_\epsilon = P_{\epsilon\lambda}$ imply

$$\dot{z}_{\epsilon\lambda}(0) = P_\lambda(y_\epsilon(0))\dot{x}_\epsilon(0) + O(\epsilon) \to P_\lambda(x_*)v_*.$$

Thus, the kinetic energy converges at $t = 0$,

$$T_{\epsilon\lambda}^{\perp}(0) \to \tfrac{1}{2}\langle P_\lambda(x_*)v_*, P_\lambda(x_*)v_*\rangle,$$

which finishes the proof of the lemma.   □

---

[32]Here is the second of the two places mentioned earlier, where a metric-dependent term drops out like magic because of the weak virial theorem. Again, the term would be trivially zero for a flat manifold, for then $\dot{G}_0 = 0$.

§2.4.   Step 4: Identification of the Limit Mechanical System

In this final step of the proof of Theorem 1 we come back to the abstract limit equation of Lemma 8. In particular, we relate the time-dependent force field F_U^{hom} to the homogenized potential U_{hom} of Definition 4.

Lemma 15. *Let there be essentially no resonances of order two or three along the limit x_0. Then, the homogenized force F_U^{hom} and the homogenized potential U_{hom} are related by*

$$Q(x_0)F_U^{\text{hom}} = \text{grad}_N \, U_{\text{hom}}(x_0).$$

Proof. We work with local coordinates $(y^1, \ldots, y^n)$ of the critical submanifold N. Let $v \in T_{x_0}N$ be an arbitrary vector field along the limit x_0. By the definition of the force F_U^{hom} we get

$$\langle F_U^{\text{hom}}, v \rangle = \tfrac{1}{2} \, \text{tr} \left(\frac{\partial H(x_0)}{\partial y^i} \cdot \Sigma_0 \right) v^i.$$

Differentiating the spectral decomposition of the Hessian (II.8) yields

$$\frac{\partial H}{\partial y^i} = \sum_\lambda \frac{\partial \omega_\lambda^2}{\partial y^i} P_\lambda + \sum_\lambda \omega_\lambda^2 \frac{\partial P_\lambda}{\partial y^i}.$$

Inserting this expression, and the block-diagonalization (II.36) of Σ_0, into the force term yields

$$
\begin{aligned}
\tfrac{1}{2} \, \text{tr} \left(\frac{\partial H(x_0)}{\partial y^i} \cdot \Sigma_0 \right) &= \tfrac{1}{2} \sum_\lambda \frac{\partial \omega_\lambda^2(x_0)}{\partial y^i} \, \text{tr}(P_\lambda(x_0) \Sigma_0 P_\lambda(x_0)) \\
&\quad + \tfrac{1}{2} \sum_{\lambda\mu} \omega_\lambda^2(x_0) \, \text{tr} \left(P_\mu(x_0) \frac{\partial P_\lambda(x_0)}{\partial y^i} P_\mu(x_0) \Sigma_0 \right) \\
&= \sum_\lambda \theta_*^\lambda \frac{\partial \omega_\lambda(x_0)}{\partial y^i} = \frac{\partial U_{\text{hom}}(x_0)}{\partial y^i}.
\end{aligned}
$$

Here, we have used three facts, previously proven. First, by Eq. (II.46), that

$$\text{tr}(P_{0\lambda} \Sigma_0 P_{0\lambda}) = \sigma_\lambda = \frac{\theta_*^\lambda}{\omega_{0\lambda}}.$$

Second, by Lemma 9, that $P_\mu \cdot \partial_{y^i} P_\lambda \cdot P_\mu = 0$. Third, that the limit actions θ_*^λ are *constants*.   □

Thus, Lemma 15 implies[33] that, under the given resonance conditions, the abstract limit equation (II.30) represents the equations of motions of

---

[33] This claim follows from well-known facts on constrained motion in Lagrangian mechanics, as can be found, for instance, in [1, p. 229] and [6, p. 19].

a natural mechanical system with configuration space N. This mechanical system is given by the Lagrangian

$$\mathscr{L}_{\text{hom}}(x, \dot{x}) = \tfrac{1}{2}\langle \dot{x}, \dot{x}\rangle - V(x) - U_{\text{hom}}(x) \qquad \dot{x} \in T_x N. \qquad (\text{II.47})$$

We denote by x_{hom} the solution of this mechanical system for the initial position x_* and initial velocity $Q(x_*)v_*$.

Now, to summarize, we have shown that *if* there were essentially no resonances of order two or three along the up to now *inaccessible* limit x_0, the equality $x_0 = x_{\text{hom}}$ would hold. Surprisingly, one can decide on this equality by looking at the resonance properties of the *accessible* function x_{hom}.

Lemma 16. *If x_{hom} is non-flatly resonant up to order three, there are only finitely many resonances of order two or three along x_0, and one gets*

$$x_0 = x_{\text{hom}}.$$

Proof. In the course of this proof, the notion "resonance" will always mean a resonance of order two or three.

There are only *finitely* many resonances along x_{hom} in the *compact* time interval $[0, T]$. Otherwise we would get a converging sequence $t_j \to \bar{t}$ of crossing times for one and the same resonant surface. As a consequence, at time $\bar{t}$ there would be a *non transversal* crossing with that resonant surface. This would contradict the assumption of transversality.

Because of $x_0(0) = x_{\text{hom}}(0)$, the maximal time

$$t_* = \max\left\{ t \in [0, T] \ : \ x_0|[0, t] = x_{\text{hom}}|[0, t] \right\}$$

of equality is a well defined quantity. Suppose we have $t_* < T$. Then, there are only finitely many resonances of x_0 during the time interval $[0, t_*]$. Since x_0 and x_{hom} are C^1-functions of time, we get

$$\dot{x}_0(t_*) = \dot{x}_{\text{hom}}(t_*),$$

just using the initial values of Lemma 3 if $t_* = 0$. Hence, by the assumption on x_{hom}, if x_0 crosses a resonance surface at time t_* it does so *transversally*. As a consequence, there is a small $\delta > 0$ such that there are no further resonances of x_0 during the time interval $]t_*, t_* + \delta]$. Thus, there are essentially no resonances along x_0 during $[0, t_* + \delta]$. The summary which precedes the statement of this lemma, shows that

$$x_0|[0, t_* + \delta] = x_{\text{hom}}|[0, t_* + \delta],$$

contradicting the maximality of t_*. We therefore obtain $t_* = T$, which is equivalent to $x_0 = x_{\text{hom}}$.   □

From this lemma we conclude that the limit x_0 of the subsequence under consideration is *unique*, i.e., independent of the chosen subsequence. Thus, by Principle I.5, we may discard all extractions of subsequences and have proven the uniform convergence $x_\epsilon \to x_{\text{hom}}$, as was asserted in the statement of Theorem 1.

§3. Realization of Holonomic Constraints

Naïve intuition could lead to the expectation that, in the limit $\epsilon \to 0$, a constraining potential U just constrains the motion to the critical submanifold N. According to the Lagrange-d'Alembert principle, the limit motion would then[34] be governed by a natural mechanical system on the configuration manifold N belonging to the Lagrangian

$$\mathcal{L}_{\mathrm{con}}(x, \dot{x}) = \tfrac{1}{2}\langle \dot{x}, \dot{x} \rangle - V(x), \qquad \dot{x} \in T_x N.$$

On the other hand, physicists frequently have expressed the idea,[35] that ideal (mathematical) constraints should be thought of as being induced by appropriate "strong potentials."

From now on, the term "realization of holonomic constraints" will have the following technical meaning.

Definition 8. For a sequence $\epsilon \to 0$, let there be a family of mechanical systems given by the Lagrangian

$$\mathcal{L}_\epsilon(x, \dot{x}) = \tfrac{1}{2}\langle \dot{x}, \dot{x} \rangle - V(x) - \epsilon^{-2}U(x), \qquad \dot{x} \in T_x M.$$

The potential U is assumed to be constraining to a nondegenerate critical submanifold $N \subset M$. For fixed initial values $x_\epsilon(0) = x_* \in N$ and $\dot{x}_\epsilon(0) = v_* \in T_{x_*}M$ and a finite time interval $[0, T]$ there exists a unique sequence x_ϵ of solutions of the Euler-Lagrange equations corresponding to $\mathcal{L}_\epsilon$. Let x_{con} be the unique solution of the Euler-Lagrange equations corresponding to the holonomic constrained Lagrangian

$$\mathcal{L}_{\mathrm{con}}(x, \dot{x}) = \tfrac{1}{2}\langle \dot{x}, \dot{x} \rangle - V(x), \qquad \dot{x} \in T_x N,$$

with initial data $x_{\mathrm{con}}(0) = x_* \in N$ and $\dot{x}_{\mathrm{con}}(0) = Q(x_*)v_* \in T_{x_*}N$. If the sequence x_ϵ converges to x_{con} uniformly, $x_\epsilon \to x_{\mathrm{con}}$, one says that the potential U and the initial values (x_*, v_*) *realize the holonomic constraints* given by the submanifold N.

This definition allows statements analogously to the homogenization result, Theorem 1. The reader should notice, however, that neither a resonance conditions is employed in this definition, nor the potential U is assumed to constrain *spectrally smooth*.

We will establish conditions on both, the initial values in §3.1, and the constraining potentials in §3.2, that imply realization of constraints.

---

[34] Compare Footnote 33 on p. 49.

[35] For instance, quite controversial positions are adapted by KOPPE and JENSEN [58] and VAN KAMPEN [49]. The reader should consult the book of GALLAVOTTI [34, §3.6] for an extensive discussion of the question of "physically real" constraints. One should also look at his historical remarks in [34, §3.10].

§3.1. Conditions on the Initial Values

The following theorem completely characterizes the initial values for which a realization-of-constraints-result holds.

Theorem 2. *A pair* (x_*, v_*) *of initial values,* $x_* \in N$ *and* $v_* \in T_{x_*}M$, *realizes holonomic constraints for all potentials* U *constraining to* N, *if and only if*

$$v_* \in T_{x_*}N.$$

Proof. Let U be any potential constraining to the manifold N. We recall that during the first two steps of the proof of Theorem 1, §§2.1-2.2, neither a resonance condition has been employed, nor has been made use of the fact that U constrains *spectrally smooth*. In particular, Lemmas 6 and 8 are applicable under the current assumptions.

For the sufficiency part of the proof, let us assume that $v_* \in T_{x_*}N$. Below, we are going to prove that this implies that $\dot{z}_\epsilon \to 0$ and $\epsilon^{-1}z_\epsilon \to 0$, strongly in L^2. Hence, we obtain $\Pi_\epsilon \to 0$ and $\Sigma_\epsilon \to 0$, strongly in L^1. In particular, the weak limits are

$$\Pi_0 = \Sigma_0 = 0.$$

The force term F_U^{hom} of the abstract limit equation (II.30) vanishes, making this equation the Euler-Lagrange equation of the Lagrangian $\mathscr{L}_{\text{con}}$ with the solution $x_0 = x_{\text{con}}$—independently of the chosen subsequence that has defined the limit function x_0 of Lemma 8. By Principle I.5, this shows that we may discard the extraction of subsequences and obtain realization of constraints.

The proof of the strong convergences will be based on considering the function

$$\phi_\epsilon(t) = \int_0^t \operatorname{tr}(\Pi_\epsilon(\tau) + \Sigma_\epsilon(\tau))\, d\tau.$$

This function is intimately related to the normal energy $E_\epsilon^\perp$ of Definition 5. On the one hand, the nondegeneracy condition (II.6) implies that

$$\dot{\phi}_\epsilon(t) = \operatorname{tr}\Pi_\epsilon(t) + \operatorname{tr}\Sigma_\epsilon(t) \le 2(1 + \omega_*^{-2})E_\epsilon^\perp(t).$$

On the other hand, Lemma 6 gives the limit normal energy

$$E_0^\perp = E_* - \tfrac{1}{2}\langle \dot{x}_0, \dot{x}_0 \rangle - V(x_0).$$

Differentiating this expression and using the abstract limit equation (II.30) of Lemma 8 yields

$$\dot{E}_0^\perp = -\langle \dot{x}_0, \nabla_{\dot{x}_0}\dot{x}_0 + \operatorname{grad}V(x_0)\rangle = \langle \dot{x}_0, F_U^{\text{hom}}\rangle = \tfrac{1}{2}\langle \dot{x}_0, \operatorname{grad}H_0 : \Sigma_0\rangle.$$

By integrating in time, we obtain

$$E_0^\perp(t) = E_0^\perp(0) + \tfrac{1}{2}\int_0^t \langle \dot{x}_0(\tau), \operatorname{grad}H(x_0(\tau)) : \Sigma_0(\tau)\rangle\, d\tau. \tag{II.48}$$

From this, by considering the uniform convergence $E_\epsilon^\perp \to E_0^\perp$ and the weak* convergence $\Sigma_\epsilon \overset{*}{\rightharpoonup} \Sigma_0$, we get that correspondingly

$$E_\epsilon^\perp(t) = E_\epsilon^\perp(0) + \delta_\epsilon + \tfrac{1}{2} \int_0^t \langle \dot{x}_0(\tau), \operatorname{grad} H(x_0(\tau)) : \Sigma_\epsilon(\tau) \rangle \, d\tau, \quad t \in [0, T],$$

where $\delta_\epsilon \to 0$ for $\epsilon \to 0$. Since $\operatorname{tr} \Sigma_\epsilon = \epsilon^{-2} \langle z_\epsilon, z_\epsilon \rangle$ is a *matrix norm* for Σ_ϵ, the estimate

$$E_\epsilon^\perp(t) \le E_\epsilon^\perp(0) + \delta_\epsilon + c\phi_\epsilon(t), \qquad t \in [0, T],$$

holds for some positive constant c. Now, the initial velocity being tangential, $v_* \in T_{x_*} N$, implies that initially

$$E_\epsilon^\perp(0) = 0,$$

cf. Eq. (II.49) below. Summarizing, we end up with the differential inequality

$$\dot{\phi}_\epsilon(t) \le 2(1 + \omega_*^{-2})(\delta_\epsilon + c\phi_\epsilon(t)), \qquad t \in [0, T].$$

The Gronwall lemma yields the estimate

$$\phi_\epsilon(t) \le 2(1 + \omega_*^{-2}) \delta_\epsilon T \exp(2(1 + \omega_*^{-2})cT), \qquad t \in [0, T].$$

This reveals that $\phi_\epsilon \to 0$ uniformly, implying the asserted strong convergences in L^2.

The proof of necessity will be based on Theorem 1. Let us assume that a given pair (x_*, v_*) of initial values realizes holonomic constraints for *all* potentials U constraining to N. Now, let $\omega : N \to \mathbb{R}$ be a strictly positive, smooth function, such that

$$\operatorname{grad} \omega(x_*) \ne 0.$$

In a tubular neighborhood of N, cf. §2.1, we define a *specific* constraining potential by

$$U(x) = \tfrac{1}{2}\omega(y) \operatorname{dist}(x, N)^2, \qquad x = \exp_y z, \quad y \in N, \ z \in T_y N^\perp.$$

A simple calculation reveals that the Hessian H of U is given by

$$H(x) = \omega(x) P(x), \qquad x \in N.$$

Thus, Theorem 1 is applicable and yields the convergence $x_\epsilon \to x_0 = x_{\text{hom}}$, where x_{hom} belongs to the homogenized potential

$$U_{\text{hom}} = \theta_* \cdot \omega, \qquad \theta_* = \frac{\langle P(x_*)v_*, P(x_*)v_* \rangle}{2\,\omega(x_*)}.$$

Since by assumption $x_0 = x_{con}$, a comparison of the Euler-Lagrange equations belonging to $\mathscr{L}_{hom}$ and $\mathscr{L}_{con}$ shows that the homogenized force field must vanish at the initial position x_*,

$$0 = \operatorname{grad} U_{hom}(x_*) = \theta_* \cdot \operatorname{grad} \omega(x_*).$$

By construction of ω, this implies $\theta_* = 0$ which is equivalent to

$$P(x_*)v_* = 0, \text{ i.e., } \quad v_* \in T_{x_*}N,$$

as being asserted. $\qquad\qquad\qquad\qquad\qquad\qquad\qquad\qquad\qquad\qquad$ □

Remark 3. Theorem 2 would be just a simple corollary of the homogenization result, Theorem 1, if we made two additional assumptions; first, that the potential U under consideration constrains *spectrally smooth*, and second, that the trajectory x_{con} is non-flatly resonant up to order three. For then, the proof of sufficiency[36] would proceed as follows: An initial velocity $v_* \in T_{x_*}N$ fulfills $P_\lambda(x_*)v_* = 0$ for all indices λ. A fortiori, we get $\theta_*^\lambda = 0$ and

$$U_{hom} = 0.$$

Hence, $\mathscr{L}_{hom} = \mathscr{L}_{con}$ and Theorem 1 yields realization of constraints.

The proof of Theorem 2 provides further insight in the homogenization problem of §§1-2. In particular, it teaches that the quadratic obstructions Π_0 and Σ_0 do *not* vanish in general. Thus, we cannot modify the proof of §2 in a way which avoids weak* convergences. We summarize this aspect by stating the following Lemma.

Lemma 17. *Let the assumptions of Lemma 8 be valid. Then, for fixed initial values $x_* \in N$ and $v_* \in T_{x_*}M$ the following properties are equivalent:*

(i) $v_* \in T_{x_*}N$,

(ii) $E_\epsilon^\perp(0) = 0$,

(iii) $E_0^\perp \equiv 0$,

(iv) $\dot{z}_\epsilon \to 0$, *strongly in* L^2,

(v) $\epsilon^{-1}z_\epsilon \to 0$, *strongly in* L^2,

(vi) $\Pi_0 = 0$,

(vii) $\Sigma_0 = 0$.

---

[36]The proof of necessity has already been based on Theorem 1.

Proof. The equivalence of (i) and (ii) follows immediately from Definition 5 which gives

$$E_\epsilon^\perp(0) = T_\epsilon^\perp(0) = \tfrac{1}{2}\langle P(x_*)v_*, P(x_*)v_*\rangle. \tag{II.49}$$

The proof of Theorem 2 has shown that property (i) implies properties (iii)–(v) and that (iv) implies (vi), as well as (v) implies (vii).

Property (vii) implies the weak* convergence $\operatorname{tr}\Sigma_\epsilon \overset{*}{\rightharpoonup} 0$ and therefore

$$\epsilon^{-2}\int_0^T \langle z_\epsilon(\tau), z_\epsilon(\tau)\rangle\, d\tau = \int_0^T \operatorname{tr}\Sigma_\epsilon(\tau)\, d\tau \ \to\ 0,$$

i.e., property (v). Analogously one shows that (vi) implies (iv).

The weak virial theorem, Lemma 7, and the nondegeneracy conditions (II.4) and (II.5) prove the equivalence of (vi) and (vii). Now, by the limit relations (II.26), we obtain from (vi) and (vii) that $E_0^\perp \equiv 0$, i.e. property (iii). Finally, by using Eq. (II.49) and the uniform convergence $E_\epsilon^\perp \to E_0^\perp$ of Lemma 6, we obtain that property (iii) implies property (ii). □

§3.2. Conditions on the Constraining Potential

Here, we state a sufficient condition on the potential U which implies realization of constraints for general initial velocities $v_* \in T_{x_*}M$. Later on, in Lemma 18, we will show that this condition is also necessary, at least partially.

Theorem 3. *Let U be constraining to the critical manifold N. Suppose the spectrum $\sigma(H)$ of the Hessian H is constant on N. Then, the potential U constrains spectrally smooth and realizes holonomic constraints for all initial values $x_* \in N$, $v_* \in T_{x_*}M$.*

Proof. First, we show that U constrains spectrally smooth. Let

$$0 < \omega_*^2 \le \omega_1^2 < \ldots < \omega_s^2. \qquad s \le r,$$

denote the different, by assumption *constant*, eigenvalues of H. Denote by Γ_λ a small path in the complex plane, such that ω_λ^2 is encircled, but none of the other eigenvalues of H. The eigenprojection P_λ belonging to ω_λ^2 is given by the Cauchy integral,

$$P_\lambda(x) = \frac{1}{2\pi i}\int_{\Gamma_\lambda} (\zeta I - H(x))^{-1}\, d\zeta, \qquad x \in N,$$

as frequently used in perturbation theory, [51, Chap. II,§1.4]. Since the path Γ_λ does *not* depend on $x \in N$, and is *never* crossed by any of the constant eigenvalues of H, this representation shows the smooth dependence of $P_\lambda(x)$ on $x \in N$.

This smoothness implies that, belonging to the eigenvalues $\omega_1^2, \ldots, \omega_s^2$, each counted by multiplicity, there is a smooth field $(e_1(x), \ldots, e_r(x))$ of orthonormal bases of $T_x N^\perp$ consisting of eigenvectors of $H(x)$. Taking this *particular* basis to form the coordinates system of §2.1, we obtain a *constant* matrix representation of H,

$$H(x) = \begin{pmatrix} 0 & 0 \\ 0 & D \end{pmatrix}, \qquad D = \text{diag}(\omega_1^2, \ldots, \omega_s^2),$$

where the eigenvalues appear according to their multiplicities. Thus, the coordinate-dependent expression $\text{grad}\, H$ vanishes,

$$\text{grad}\, H = 0.$$

This implies that also the force term F_U^{hom} of the abstract limit equation (II.30) vanishes, making this equation the Euler-Lagrange equation of the Lagrangian $\mathscr{L}_{\text{con}}$ with the solution $x_0 = x_{\text{con}}$—independently of the chosen subsequence that has defined the limit function x_0 of Lemma 8. By Principle I.5, this shows that we may discard the extraction of subsequences and obtain realization of constraints. The reader should notice that Lemma 8 is valid under the assumptions made in this theorem.   □

Remark 4. Theorem 3 would be just a simple corollary of the homogenization result, Theorem 1, if we made the additional assumption, that the constant normal frequencies ω_λ satisfy no resonance of order three,[37]

$$\omega_\lambda + \omega_\mu \neq \omega_\nu$$

for all indices λ, μ, and ν. For then, the proof would proceed as follows: Because all the normal frequencies are *constant*, we would get $U_{\text{hom}} = \sum_\lambda \theta_*^\lambda \omega_\lambda = \text{const}$. Hence, the Lagrangians $\mathscr{L}_{\text{hom}}$ and $\mathscr{L}_{\text{con}}$ differ only by a constant. Thus, they have the same Euler-Lagrange equations and Theorem 1 gives realization of constraints.

Example 3. A simple example for a potential U which satisfies the conditions of Theorem 3, is provided by

$$U(x) = \tfrac{1}{2} \text{dist}(x, N)^2.$$

An easy calculation reveals that the Hessian of this function is just the projection onto the normal bundle,

$$H = P.$$

Thus, there is only one normal frequency, $\omega_1 \equiv 1$. Theorem 3 states realization of constraints for all initial values $x_* \in N$, $v_* \in T_{x_*} M$. Notice that the simple proof mentioned in Remark 4 is applicable here.

---

[37]Notice that there are no resonance of order two. For the normal frequencies being constant, we can discard *identical* ones by combining corresponding eigenspaces of H, cf. §1.3.

Now, we state a partial converse of Theorem 3. Its proof relies on the homogenization result, Theorem 1.

Lemma 18. *Let U constrain spectrally smooth to the critical manifold N. Suppose, that the normal frequencies have no resonances of order two or three. Let U realize holonomic constraints for all initial values $x_* \in N$ and $v_* \in T_{x_*} M$. Then, the spectrum $\sigma(H)$ of the Hessian H is constant on N.*

Proof. Let us fix an initial position $x_* \in N$. For a given index λ, we can choose an initial velocity $v_* \in T_{x_*} M$ such that

$$\theta_*^\lambda = 1, \qquad \theta_*^\mu = 0 \quad \mu \neq \lambda.$$

By the assumptions and Theorem 1, the homogenized force field must vanish at the initial position x_*,

$$0 = \operatorname{grad} U_{\mathrm{hom}}(x_*) = \operatorname{grad} \omega_\lambda(x_*).$$

Otherwise, the limit x_{hom} would not satisfy the Euler-Lagrange equation belonging to the Lagrangian $\mathcal{L}_{\mathrm{con}}$ of holonomic constraints. Since x_* and λ are arbitrary, each normal frequency, and therefore the spectrum of H, is constant on the critical submanifold N.   □

In §3.1, Lemma 17, we have shown that the conditions of Theorem 2 can be characterized by a normal energy $E_\epsilon^\perp$ that vanishes in the limit, $E_0^\perp \equiv 0$. Partially, a similar result holds here.

Corollary 1. *Let the assumptions of Theorem 3 be valid. Then, the normal energy $E_\epsilon^\perp$ is an adiabatic invariant,[38] i.e.,*

$$E_\epsilon^\perp \to E_0^\perp = \mathrm{const},$$

uniformly in $t \in [0, T]$.

Proof. By the proof of Theorem 3, there is a coordinate system such that $\operatorname{grad} H = 0$. Now, the adiabatic invariance follows from Lemma 6 and the integral representation (II.48) of $E_0^\perp$.   □

The converse is not true, since one can easily construct homogenized potentials U_{hom} such that the initial values $(x_*, Q(x_*)v_*)$ are fixed points. This would lead to $E_0^\perp = \mathrm{const}$ without a realization-of-constraint-result being valid.

§3.3.   Bibliographical Remarks

The first mathematical proof of the sufficiency part of Theorem 2 was given in 1957 by RUBIN and UNGAR [82]. These authors consider, however, the Euclidean space $M = \mathbb{R}^m$ only. For this case, the theorem has been proven

---

[38]Compare Footnote 31 on p. 46.

by BORNEMANN and SCHÜTTE [18, Theorem 3.1] along the lines presented here. The specific case of codimension $r = 1$ has been discussed in the work of the physicists KOPPE and JENSEN [58] and VAN KAMPEN [49]. This case also appears in form of an example in a textbook of ARNOLD [5, Chap. 17A]. For *general* Riemannian manifolds, the sufficiency part is stated as Theorem 9 in the encyclopedic survey of classical mechanics, [6, Chap. 1,§6.2] by ARNOLD, KOZLOV, and NEISHTADT. However, for a proof they refer to the work of RUBIN and UNGAR [82], and of TAKENS [94]. On the latter work a proof could only be based along the lines of Remark 3. In addition, one would have to exclude *any* resonances of order two or three.

The necessity part of Theorem 2 appears to be entirely new. It relies on our generalization, Theorem 1, of the work of TAKENS [94], which has made possible to consider eigenvalues of multiplicity r for the first time.

In the setting of Theorem 2, the convergences in normal direction are actually *strong*, cf. Lemma 17. This strong convergence allows to apply the "slow manifold" technique of KREISS [59], which has been worked out 1993 by LUBICH [66, Theorem 2.2] in his work on Runge-Kutta methods for "stiff" mechanical systems.

For the Euclidean space $M = \mathbb{R}^m$, Theorem 3 has been stated and proved 1983 by GALLAVOTTI [34, Chap. 3, §3.8] as "Arnold's theorem" in view of a remark, which was made by ARNOLD on p. 91f of his textbook [5]. TAKENS [94, p. 429] offers a proof along the lines of Remark 4, however, under the more restrictive assumption that the normal frequencies satisfy *no* resonances of order two or three, nor have any multiplicity bigger than one. Thus, his proof cannot even handle Example 3, which requires the multiplicity r. For codimension $r = 1$, one can find a discussion in the work of RUBIN and UNGAR [82], KOPPE and JENSEN [58], and VAN KAMPEN [49]. A proof of Gallavotti's theorem along the lines presented here was given by BORNEMANN and SCHÜTTE [18, Theorem 3.2]. In 1993, SCHMIDT [86, Prop. 3] has generalized the work of Gallavotti to Riemannian manifolds and nonconservative forces. This author refers to Theorem 3 as the "Arnold-Gallavotti theorem."

Under the assumptions[39] of Theorem 3, BENETTIN, GALGANI, and GIORGILLI [11][12] have shown the adiabatic invariance of the normal energy $E_\epsilon^\perp$ for *exponentially large* times. To be precise, these authors have proven a Nekhoroshev-type of result,

$$|E_\epsilon^\perp(t) - E_\epsilon^\perp(0)| < \epsilon \qquad \text{for} \quad 0 \le t \le \exp(b\epsilon^{-a}),$$

where a and b are positive constants. In general, one has $a = 1/r$, where r denotes the codimension of the critical manifold N; but for instance, the specific potential of Example 3 yields $a = 1$ in any dimension. These results should be contrasted with the comparatively rather trivial Corollary 1 and the estimate given by SCHMIDT [86, Prop. 1].

---

[39]For *holomorphic* potentials and $M = \mathbb{R}^m$ Euclidean.

There is an intimate connection of the realization-of-constraint problem to various singular limits in fluid dynamics, such as the incompressible
limit of compressible fluid flow, or the quasi-geostrophic limit of geophysical fluid flow. The condition $v_* \in T_{x_*} N$ of Theorem 2 corresponds to initial
data which are *balanced*. For balanced data, a Hamiltonian approach to
the incompressible limit was given by EBIN [26], a careful perturbation
analysis can be found in the work of KLAINERMAN and MAJDA [55], and
weak topologies are considered by CHEMIN [21]. Later on, it has turned
out that the "constraining" potentials of these flow problems satisfy conditions that correspond to the one given in Theorem 3. Thus, in general,
unbalanced data lead to the same limit equations as balanced data. For
the incompressible limit consult the work of SCHOCHET [87], for the quasi-
geostrophic limit EMBID and MAJDA [27].

§4.   Spectrally Nonsmooth Constraining Potentials: Takens Chaos

According to Theorem 1, the limit dynamics of a natural mechanical system with a strong, smoothly constraining potential U is governed by the
homogenized Lagrangian $\mathscr{L}_{\mathrm{hom}}$. This Lagrangian does *only* depend on the
fixed initial position x_* and on the *limit* initial velocity v_*. In particular,
the details of the limiting process $\dot{x}_\epsilon(0) \to v_*$ do not matter at all—which
is a far-reaching *stability property* of the homogenization process.

A completely different situation can be encountered if the constraining
potential U *fails* to constrain spectrally smooth. This was discovered by
TAKENS [94] who sketched how to construct generic[40] examples for which
the limit set is not uniquely determined by the limit initial values (x_*, v_*).
We will construct an *explicit* example which shows up the strange properties
predicted by TAKENS.

The arguments of §1.7 show that a constraining potential U generically fails to constrain spectrally smooth if the normal frequencies have
a *generic* resonance of codimension two. Thus, we need at least a two-
parameter dependence of the Hessian, i.e., we have the lower bound $n \geq 2$
for the dimension of the critical manifold. On the other hand, for having
eigenvalue-resonances at all, the Hessian has to operate on a normal space
that is two-dimensional at least, i.e., $r \geq 2$.

We will construct a minimal example with $n = 2$, $r = 2$.

On the Euclidean space $\mathbb{R}^4$, with coordinates $x = (y^1, y^2, z^1, z^2)$, we
consider the singularly perturbed natural mechanical system given by the
Lagrangian

$$\mathscr{L}_\epsilon = \tfrac{1}{2}|\dot{y}|^2 + \tfrac{1}{2}|\dot{z}|^2 - \epsilon^{-2}U(y, z).$$

---

[40]Here, the term "generic" is understood with respect to arbitrary smooth perturbations of the potential.

The potential U is given by[41]

$$U(y,z) = \tfrac{1}{2}\langle H(y)z, z\rangle, \quad H(y) = \tfrac{1}{4}\left(I + \begin{pmatrix} y^1 & y^2 \\ y^2 & -y^1 \end{pmatrix}\right).$$

The eigenvalues of H are

$$\omega_1^2(y) = \tfrac{1}{4}(1 - |y|), \qquad \omega_2^2(y) = \tfrac{1}{4}(1 + |y|).$$

Thus, if we restrict ourselves, for instance, to the cylinder

$$M = \{(y, z) : |y| \le 0.95\},$$

the matrix H is uniformly positive definite and U constrains to the critical submanifold

$$N = \{x = (y, z) \in M : z = 0\}.$$

The spectral decomposition of H is best described using polar coordinates,

$$y^1 = r\cos\phi, \qquad y^2 = r\sin\phi.$$

The eigenvectors of H which belong to ω_1^2 and ω_2^2 are then given by

$$\begin{pmatrix} -\sin(\phi/2) \\ \cos(\phi/2) \end{pmatrix}, \quad \text{resp.} \quad \begin{pmatrix} \cos(\phi/2) \\ \sin(\phi/2) \end{pmatrix}.$$

The occurrence of the argument $\phi/2$ shows that these eigenvectors are defined up to a sign only. For a unique representation we have to cut the y-plane along a half-axis, e.g., along $\phi = 3\pi/2$. Hence, we restrict the angular variable to the open interval

$$\phi \in \left]-\frac{\pi}{2}, \frac{3\pi}{2}\right[.$$

This way, the eigenvectors become smooth vector fields uniquely defined on the cut plane

$$\mathbb{R}_c^2 = \mathbb{R}^2 \setminus \{y : y^1 = 0, \ y^2 \le 0\}.$$

They cannot, however, be continued beyond the cut, but instead change their mutual roles there. Thus, there is *no* neighborhood of $x = 0$ where the potential U is constraining *spectrally smooth*.

We consider the following family of initial values depending on a parameter $\mu \ge 0$:

$$y_\epsilon(0) = (9/16, 0), \qquad \dot{y}_\epsilon(0) = (0, \mu), \qquad z_\epsilon(0) = (0, 0), \qquad \dot{z}_\epsilon(0) = (1, 0).$$

For the discussion of the singular limit $\epsilon \to 0$ we have to distinguish *two cases*.

---

[41]The matrix $\begin{pmatrix} y^1 & y^2 \\ y^2 & -y^1 \end{pmatrix}$ is the famous example of RELLICH for a smooth symmetric matrix that is not smoothly diagonalizable, cf. [80, §2][51, Chap. 2, Example 5.12]. In fact, the arguments of Appendix A show that this matrix inevitably enters as a kind of normal form for any resonance of codimension two.

The Case $\mu = 0$. In this case, the limit trajectory $y_0^{\mu=0}$ crosses the singularity $y = 0$ of the spectral decomposition of H.

Lemma 19. *Let $\mu = 0$ and $T = 11\sqrt{5}/6$. For ϵ small enough, we obtain*

$$y_\epsilon^2 \equiv z_\epsilon^2 \equiv 0, \qquad |y_\epsilon^1(t)| \leq 0.95 \quad \forall t \in [0, T].$$

There are the convergences $z_\epsilon^1 = O(\epsilon)$ and

$$y_\epsilon^1 \to \eta_{\mu=0} \quad in \quad C^1[0, T],$$

where $\eta_{\mu=0}$ is the solution of the initial value problem

$$\ddot{\eta} = -\frac{1}{5\sqrt{1+\eta}}, \qquad \eta(0) = 9/16, \quad \dot{\eta}(0) = 0. \qquad (\text{II}.50)$$

The limit function crosses the singularity $y = 0$ at time $t_* = 7\sqrt{5}/6$,

$$\eta(t_*) = 0.$$

Proof. The equations of motion are given by

$$\ddot{y}_\epsilon^1 = -\tfrac{1}{8}\epsilon^{-2}\left((z^1)^2 - (z^2)^2\right), \qquad \ddot{y}_\epsilon^2 = -\tfrac{1}{4}\epsilon^{-2}z^1 z^2,$$

and

$$\ddot{z}_\epsilon^1 = -\tfrac{1}{4}\epsilon^{-2}\left(z^1 + (y^1 z^1 + y^2 z^2)\right), \qquad \ddot{z}_\epsilon^2 = -\tfrac{1}{4}\epsilon^{-2}\left(z^2 + (y^2 z^1 - y^1 z^2)\right).$$

Given the initial values with $\mu = 0$, one readily observes that $y_\epsilon^2 \equiv z_\epsilon^2 \equiv 0$ and the system reduces to

$$\ddot{y}_\epsilon^1 = -\tfrac{1}{8}\epsilon^{-2}(z^1)^2, \qquad \ddot{z}_\epsilon^1 = -\tfrac{1}{4}\epsilon^{-2}(1 + y^1)z^1.$$

In the (y^1, z^1)-space, these are the equations of motion for the natural mechanical system

$$\widehat{\mathscr{L}}_\epsilon = \tfrac{1}{2}|\dot{y}^1|^2 + \tfrac{1}{2}|\dot{z}^1|^2 - \epsilon^{-2}\hat{U}(y^1, z^1), \qquad \hat{U}(y^1, z^1) = \tfrac{1}{8}(1 + y^1)(z^1)^2.$$

As long as $|y^1| \leq 0.95$, the potential $\hat{U}$ is constraining spectrally smooth to the critical submanifold $z^1 = 0$ of codimension *one*. According to Definition 4, we obtain

$$w(y^1) = \tfrac{1}{2}\sqrt{1 + y^1}, \qquad \theta_* = \tfrac{4}{5}, \qquad \hat{U}_{\text{hom}}(y^1) = \tfrac{2}{5}\sqrt{1 + y^1}.$$

Corresponding to this homogenized potential $\hat{U}_{\text{hom}}(y^1)$, the Newtonian equation of motion is given by Eq. (II.50). Denote its solution by η. By conservation of energy we get the first order equation

$$\dot{\eta} = -\sqrt{1 - \tfrac{4}{5}\sqrt{1+\eta}}.$$

Using separation of variables, the time t_* at which η equals zero is given by

$$t_* = \int_0^{9/16} \frac{d\eta}{\sqrt{1 - \frac{4}{5}\sqrt{1+\eta}}} = \tfrac{7\sqrt{5}}{6} = 2.6087\ldots.$$

Accordingly, the time T at which there is $\eta(T) = -15/16 = -0.9375$ is given by

$$T = \int_{-15/16}^{9/16} \frac{d\eta}{\sqrt{1 - \frac{4}{5}\sqrt{1+\eta}}} = \tfrac{11\sqrt{5}}{6} = 4.0994\ldots.$$

Now, Theorem 1 is applicable for small ϵ and yields the claimed strong convergence $y_\epsilon^1 \to \eta$ in $C^1[0,T]$. Lemma 2 gives $z_\epsilon^1 = O(\epsilon)$.   □

We will denote the limit trajectory in y-space by $y_0^{\mu=0} = (\eta_{\mu=0}, 0)$.

The Case $\mu > 0$. In this case, the y-components stay in the cut plane $\mathbb{R}_c^2$, the potential U is constraining spectrally smooth, and Theorem 1 is directly applicable.

Lemma 20. *Let $\mu > 0$ be small enough and $T = 11\sqrt{5}/6$. For ϵ small enough, we obtain*

$$0 < \mu \le |y_\epsilon^1(t)| \le 0.95 \quad \forall t \in [0,T],$$

and the values of y_ϵ stay in the cut plane $\mathbb{R}_c^2$. There are the convergences $z_\epsilon = O(\epsilon)$ and

$$y_\epsilon \to y_0 \quad in \quad C^1([0,T], \mathbb{R}^2),$$

where y_0 is the solution of the initial value problem

$$\ddot{y}_0 = -\frac{1}{5\sqrt{1+|y_0|}} \frac{y_0}{|y_0|}, \qquad y_0(0) = (9/16, 0), \quad \dot{y}_0(0) = (0, \mu). \quad \text{(II.51)}$$

Proof. For values $y \in \mathbb{R}_c^2$ with $|y| \le 0.95$, the potential U is constraining spectrally smooth. According to Definition 4, the given initial values yield

$$\theta_*^1 = 0, \qquad \theta_*^2 = \tfrac{4}{5}, \qquad U_{\text{hom}}(y) = \theta_*^2\, \omega_2(y) = \tfrac{2}{5}\sqrt{1+|y|}.$$

Thus, the homogenized mechanical system is given by the Lagrangian

$$\mathscr{L}_{\text{hom}} = \tfrac{1}{2}|\dot{y}|^2 - \tfrac{2}{5}\sqrt{1+|y|} = \tfrac{1}{2}\dot{r}^2 + \tfrac{1}{2}r^2\dot{\phi}^2 - \tfrac{2}{5}\sqrt{1+r}.$$

The Euler-Lagrange equations (II.51) transform into polar coordinates as

$$\ddot{r} = r\dot{\phi}^2 - \frac{1}{5\sqrt{1+r}}, \qquad \frac{d}{dt}\left(r^2\dot{\phi}\right) = 0,$$

with the initial values $r(0) = 9/16$, $\phi(0) = 0$, $\dot{r}(0) = 0$, and $\dot{\phi}(0) = 16\mu/9$. We obtain

$$\dot{\phi} = \tfrac{9}{16}\mu r^{-2} > 0,$$

showing that ϕ is strictly monotonely increasing. Elimination of the cyclic variable ϕ and conservation of energy yields the first order equation

$$\dot{r}^2 + \tfrac{81}{256}\mu^2 r^{-2} + \tfrac{4}{5}\sqrt{1+r} = 1 + \mu^2.$$

Thus, for μ small enough, there is a periodic motion between the two extrema

$$r_0 = \tfrac{9\sqrt{5}}{16}\mu + O(\mu^2) = \mu \cdot 1.2577\ldots + O(\mu^2), \qquad r_1 = \tfrac{9}{16} = 0.5625.$$

By separation of variables, the period Δt of the motion is given by

$$\Delta t = 2\int_{r_0}^{r_1} \frac{dr}{\sqrt{1+\mu^2 - \tfrac{4}{5}\sqrt{1+r} - \tfrac{81}{256}\tfrac{\mu^2}{r^2}}}$$

$$\to 2\int_0^{9/16} \frac{dr}{\sqrt{1 - \tfrac{4}{5}\sqrt{1+r}}} = 2t_*, \qquad \mu \to 0.$$

During that period the angular variable ϕ increases for the amount $\Delta\phi$, given by

$$\Delta\phi = 2\int_{r_0}^{r_1} \frac{\dot{\phi}\,dr}{\dot{r}} = 2\int_{r_0}^{r_1} \frac{9\mu\,dr}{16r^2\sqrt{1+\mu^2 - \tfrac{4}{5}\sqrt{1+r} - \tfrac{81}{256}\tfrac{\mu^2}{r^2}}}$$

$$= 2\int_{\frac{r_0}{r_1}}^{1} \frac{9\mu\,dz}{16r_0\sqrt{1+\mu^2 - \tfrac{4}{5}\sqrt{1+r_0/z} - \tfrac{81}{256}\tfrac{\mu^2 z^2}{r_0^2}}}$$

$$\to 2\int_0^1 \frac{dz}{\sqrt{1-z^2}} = \pi, \qquad \mu \to 0.$$

Thus, for μ small enough, the solution of the homogenized system stays well inside the cut plane $\mathbb{R}_c^2$ for $[0,T]$. Theorem 1 and Lemma 2 are applicable and show the asserted convergences. These in turn imply the estimates for the trajectories y_ϵ themselves. $\qquad\square$

If one takes the limit $\mu \downarrow 0$ of the ϵ-limit solution y_0 given by Eq. (II.51) one ends up with the function $y_0^{\mu\downarrow0} = (\eta_{\mu\downarrow0}, 0)$, where $\eta_{\mu\downarrow0}$ fulfills the differential equation

$$\ddot{\eta} = \begin{cases} -\dfrac{1}{5\sqrt{1+\eta}} & \eta > 0, \\[2mm] \dfrac{1}{5\sqrt{1-\eta}} & \eta < 0, \end{cases}$$

with the initial values $\eta(0) = 9/16$, $\dot{\eta}(0) = 0$.

Discussion. The different limit behavior of the two cases can be explained roughly as follows. For $\mu > 0$ we are in the situation of a spectrally smooth constraining potential. Thus, eigenvalues and eigenspaces are followed according to their *number*, and even for $\mu \downarrow 0$ the dynamics is governed by the potential

$$U_{\text{hom}}(y) = \theta_*^1 \, \omega_1(y) + \theta_*^2 \, \omega_2(y),$$

and the initial values

$$y_0(0) = (9/16, 0), \qquad \dot{y}_0(0) = (0, 0).$$

On the other hand, for $\mu = 0$ we are in a dimensionally reduced situation and follow the active eigenvalue *smoothly*. This means we have to *change* its number after passing $y = 0$. Here, the dynamics is governed by the potential

$$\hat{U}_{\text{hom}}(y^1, 0) = \begin{cases} \theta_*^1 \, \omega_1(y^1, 0) + \theta_*^2 \, \omega_2(y^1, 0) & y^1 \geq 0, \\[2mm] \theta_*^2 \, \omega_1(y^1, 0) + \theta_*^1 \, \omega_2(y^1, 0) & y^1 \leq 0, \end{cases}$$

according to the same initial values

$$y_0(0) = (9/16, 0), \qquad \dot{y}_0(0) = (0, 0).$$

Hence, as long as the singularity $y = 0$ does not appear, Theorem 1 is applicable and the two limit functions are equal,

$$y_0^{\mu \downarrow 0}(t) = y_0^{\mu = 0}(t), \qquad t \in [0, t_*].$$

However, after passing the singularity $y = 0$ at $t = t_*$, the limits $\epsilon \to 0$ and $\mu \downarrow 0$ are *not* longer interchangeable,

$$\lim_{\epsilon \to 0} \lim_{\mu \downarrow 0} y_\epsilon(t) = y_0^{\mu = 0}(t) \neq y_0^{\mu \downarrow 0}(t) = \lim_{\mu \downarrow 0} \lim_{\epsilon \to 0} y_\epsilon(t), \qquad t \in]t_*, T].$$

This non-commutativity is illustrated in Figure II.1. Now, if we consider the simultaneous limit by taking an ϵ-dependent sequence $\mu(\epsilon) \downarrow 0$, the resulting limit solution y_0 would depend on *how* the limit initial velocity

$$\lim_{\epsilon \to 0} \dot{y}_\epsilon(0) = \lim_{\epsilon \to 0} (0, \mu(\epsilon)) = (0, 0)$$

is obtained.

This is in sharp contrast to the assertion of Theorem 1, showing the *necessity* of potentials which are constraining *spectrally smooth*. The situation here is even worse: By continuity we may obtain as the limit value of $y_\epsilon(t)$ at time $t > t_*$ *any* value $\tilde{y} = (\tilde{\eta}, 0)$ with

$$\eta_{\mu=0}(t) \leq \tilde{\eta} \leq \eta_{\mu \downarrow 0}(t).$$

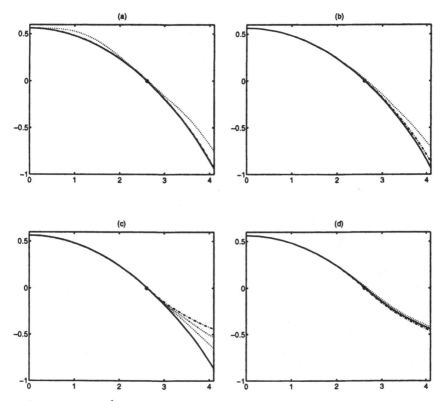

Legend: The y^1-components of the following trajectories are shown versus time:

(a) Trajectories for $\mu = 0$ and different ϵ: the limit $\epsilon = 0$ (solid), $\epsilon = 0.1$ (dashed), and $\epsilon = 0.5$ (dotted)

(b) Trajectories for $\epsilon = 0.1$ and different μ: $\mu = 0$ (solid), $\mu = 0.05$ (dashed), and $\mu = 0.1$ (dotted)

(c) Trajectories for $\mu = 0.05$ and different ϵ: $\epsilon = 0.1$ (solid), $\epsilon = 0.01$ (dotted), $\epsilon = 0.005$ (dotted), and the limit $\epsilon = 0$ (dashed)

(d) The limit $\epsilon = 0$ for $\mu > 0$ and different μ: $\mu = 0.05$ (solid), the limit $\mu \downarrow 0$ (dashed), and $\mu = 0.1$ (dotted)

Notice that the dashed line of a subfigure appears as the solid line in the next subfigure.

Figure II.1: Illustration of the non-commutativity of the limits $\mu \to 0$ and $\epsilon \to 0$

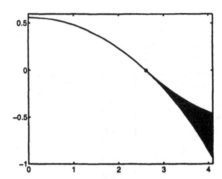

Figure II.2: Illustration (y^1 vs. t) of a funnel as the limit set.

One just has to choose the sequence $\mu(\epsilon)$ accordingly.

Until the impact time $t = t_*$ the limit dynamics is governed by the homogenized potential U_{hom}, afterwards it changes *suddenly* in a completely unpredictable way—given the limit initial values $x_* = (9/16, 0)$ and $v_* = (0,0)$ as the only data. With these data only, we have to regard the *funnel*[42] between the two extreme cases $\eta_{\mu=0}$ and $\eta_{\mu\downarrow0}$ as the *limit set*. Figure II.2 illustrates the situation. With regard to TAKENS' discovery of this effect, we speak of *Takens chaos*, cf. [18][57].

Because the strong influence of the particularities of the limiting process $\dot{x}_\epsilon(0) \to v_*$ shows up in time with a considerable *delay*, it should be clear that the structure of Theorem 1 is completely destroyed. There is no hope for a comparably elegant result for spectrally *nonsmooth* constraining potentials.

§5.   Generalization: Friction and Gyroscopic Forces

Up to now, we have only studied weak forces[43] that were belonging to a potential V. As a last step in this chapter, we intend to allow for more general force fields. Therefore, we consider the dynamics of a mechanical system that is governed by the Lagrangian

$$\mathscr{L}_\epsilon(x, \dot{x}) = \tfrac{1}{2}\langle \dot{x}, \dot{x}\rangle - V(x) - \epsilon^{-2}U(x), \qquad \dot{x} \in T_x M,$$

subject to an *external force* of the form

$$F(x, \dot{x}) \in T_x M, \qquad \dot{x} \in T_x M.$$

---

[42]The appearance of funnels bears similarities with non-uniquely solvable initial value problems, cf. [48][56][74].

[43]I.e., forces that are not responsible for the singular perturbation.

Instead of Eq. (II.1), the equations of motion are now—by definition—given as

$$\nabla_{\dot{x}_\epsilon} \dot{x}_\epsilon + \operatorname{grad} V(x_\epsilon) + \epsilon^{-2} \operatorname{grad} U(x_\epsilon) = F(x_\epsilon, \dot{x}_\epsilon). \qquad (II.52)$$

However, we cannot attack the problem of homogenization in quite that generality.

If the energy arguments of §2.1 apply, we will get the convergences

$$x_\epsilon \to x_0, \qquad \dot{x}_\epsilon \overset{*}{\rightharpoonup} \dot{x}_0.$$

Thus, the force F, depending on the just weakly* converging velocities, will have a nontrivial impact on the limit dynamics. For our method to work this impact should only appear on the level of energies, but not on the level of forces. This means, for obtaining an abstract limit equation analogous to the one given in Lemma 8, we have to require the weak* continuity

$$F(x_\epsilon, \dot{x}_\epsilon) \overset{*}{\rightharpoonup} F(x_0, \dot{x}_0).$$

The most general force that guarantees this weak* continuity is *affine* in the velocities, [23, Thm. I.1.1]:

$$-F(x, \dot{x}) = F_0(x) + K(x) \cdot \dot{x},$$

where $K : TM \to TM$ is a field of linear operators.

The energy of the system (II.52), again a *constant* of motion, is given by the expression

$$E_\epsilon = \tfrac{1}{2} \langle \dot{x}_\epsilon, \dot{x}_\epsilon \rangle + V(x_\epsilon) + \epsilon^{-2} U(x_\epsilon) - \int_0^t \langle F(x_\epsilon, \dot{x}_\epsilon), \dot{x}_\epsilon \rangle \, d\tau. \qquad (II.53)$$

For the energy arguments of §2.1 to apply, the last term has to be bounded from below, independent of the specification of x_ϵ. This way we obtain further restrictions of the force term F. Quite the most general admissible form is given as follows:[44]

- F_0 is a potential force, belonging to a potential that is bounded from below,

- the linear operator K is *nonnegative*, i.e., $\langle K(x)v, v \rangle \geq 0$ for all tangential vectors $v \in T_x M$.

---

[44]To be specific: For all $x_* \in M$, $v_* \in T_{x_*} M$ and times T there has to be a constant $\beta = \beta(x_*, v_*, T)$ such that

$$-\int_0^t \langle F(x(\tau), \dot{x}(\tau)), \dot{x}(\tau) \rangle \, d\tau \geq \beta, \qquad 0 \leq t \leq T,$$

where $x : [0, T] \to M$ is any given smooth path starting as $x(0) = x_*$, $\dot{x}(0) = v_*$. This readily implies the nonnegativity of K. Moreover, if K is skew-adjoint with respect to the Riemannian metric, F_0 has *necessarily* to be a potential force, belonging to a potential that is bounded from below.

We may put $F_0 = 0$ by incorporating the corresponding potential into the weak potential V of the Lagrangian.

We split the nonnegative operator $K = A + S$ into its nonnegative selfadjoint part A and its skew-adjoint part S,

$$A = \tfrac{1}{2}(K + K^*) \geq 0, \qquad S = \tfrac{1}{2}(K - K^*).$$

Correspondingly, there is the splitting

$$F(x, \dot{x}) = F_{\text{fric}}(x, \dot{x}) + F_{\text{gyro}}(x, \dot{x})$$

of the force $F(x, \dot{x}) = -K(x)\dot{x}$ into *viscous friction* $F_{\text{fric}}(x, \dot{x}) = -A(x)\dot{x}$, defined by $A = A^* \geq 0$, and a *gyroscopic force* $F_{\text{gyro}}(x, \dot{x}) = -S(x)\dot{x}$, defined by $S = -S^*$. This way, we take the two major classes of velocity-dependent forces into account that are of importance in applications.

For the sake of simplicity of the result, we restrict ourselves to a specific class of viscous friction.

Definition 9. A force field $F_{\text{fric}}(x, \dot{x}) = -A(x)\dot{x}$ $(A = A^* \geq 0)$ of viscous friction is called κ-*isotropic* normal to the submanifold $N \subset M$ if there is some real constant $\kappa \geq 0$ such that $\langle A(x)v, v \rangle = \kappa \langle v, v \rangle$ for all $x \in N$ and $v \in T_x N^\perp$.

Using the projection P, we can rewrite the κ-isotropy as

$$P(x)A(x)P(x) = \kappa \cdot P(x), \qquad x \in N. \tag{II.54}$$

Now, Theorem 1 generalizes to systems with κ-isotropic friction and gyroscopic forces as follows.

Theorem 4. *For a sequence $\epsilon \to 0$, consider the family of mechanical systems given by the Lagrangian*

$$\mathscr{L}_\epsilon(x, \dot{x}) = \tfrac{1}{2}\langle \dot{x}, \dot{x} \rangle - V(x) - \epsilon^{-2}U(x), \qquad \dot{x} \in T_x M,$$

subject to the external forces $F_{\text{fric}}(x, \dot{x}) + F_{\text{gyro}}(x, \dot{x})$. The potential U is assumed to constrain spectrally smooth to a nondegenerate critical submanifold $N \subset M$ and the force field F_{fric} of viscous friction to be κ-isotropic normal to N. Let the initial positions be fixed on the critical submanifold, $x_\epsilon(0) = x_* \in N$, and the initial velocities convergent in $T_{x_*}M$, $\dot{x}_\epsilon(0) \to v_* \in T_{x_*}M$. Then, for a finite time interval $[0, T]$, there exists a unique sequence x_ϵ of solutions of the equations of motion.

Let U_{hom} be the homogenization of U with respect to the limit initial values (x_*, v_*), as given by Definition 4. We denote by x_{hom} the unique solution of the equations of motion corresponding to the non-autonomous Lagrangian

$$\mathscr{L}_{\text{hom}}^\kappa(x, \dot{x}, t) = \tfrac{1}{2}\langle \dot{x}, \dot{x} \rangle - V(x) - e^{-\kappa t}U_{\text{hom}}(x), \qquad \dot{x} \in T_x N,$$

subject to the projected external forces $Q(x)F_{\text{fric}}(x, \dot{x}) + Q(x)F_{\text{gyro}}(x, \dot{x})$
and initial values $x_{\text{hom}}(0) = x_* \in N$ *and* $\dot{x}_{\text{hom}}(0) = Q(x_*)v_* \in T_{x_*}N$.

If x_{hom} *is non-flatly resonant up to order three, the sequence* x_ϵ *converges uniformly to* x_{hom} *on* $[0, T]$.

Proof. The proof may be obtained by slightly modifying the proof of Theorem 1 as given in §2. We will just highlight these modifications; all the other results of §2 remain valid with literally the same proof.

Step 1: Equi-Boundedness (as in §2.1). The energy (II.53) of the system specifies as

$$E_\epsilon = \tfrac{1}{2}\langle \dot{x}_\epsilon, \dot{x}_\epsilon \rangle + V(x_\epsilon) + \epsilon^{-2}U(x_\epsilon) + \int_0^t \langle K(x_\epsilon)\dot{x}_\epsilon, \dot{x}_\epsilon \rangle \, d\tau.$$

Because $K(x)$ is a nonnegative linear operator, the integral is always nonnegative.

Step 2: The Weak Virial Theorem (as in §2.2). First, we note that the splitting of energy, as given in Lemma 6, has to be defined differently. However, this modification leaves the definition of the energies of normal motion, $E_\epsilon^\perp = T_\epsilon^\perp + U_\epsilon^\perp$, *untouched.* Since Lemma 6 has no impact on the proof other than justifying the *notion* of energy components, we postpone the detailed discussion of the modified energy splitting.

Second, the abstract limit equation (II.30) of Lemma 8 changes to

$$\ddot{x}_0 + \Gamma(x_0)(\dot{x}_0, \dot{x}_0) + F_V(x_0) + F_U^{\text{hom}}(t) - F(x_0, \dot{x}_0) \perp T_{x_0}N. \quad \text{(II.55)}$$

Here, we have made use of $F(x, \dot{x}) = -K(x)\dot{x}$ being *linear* in the velocity argument.

Step 3: Adiabatic "Invariance" of the Normal Actions (as in §2.3). The detailed resolution of the right-hand-side of the componentwise oscillator equations (II.45), as given by Lemma 12, *is* affected by the velocity-dependence of the external force term $F(x, \dot{x})$. Because of

$$P_{\epsilon\lambda}F(x_\epsilon, \dot{x}_\epsilon) \equiv -P_{\epsilon\lambda}K_\epsilon \dot{z}_\epsilon \quad (\text{mod } C^0\text{-lim})$$

there appears an additional term, leading to the modified equation

$$\ddot{z}_{\epsilon\lambda}^i + \epsilon^{-2}\omega_{\epsilon\lambda}^2 z_{\epsilon\lambda}^i = -P_{\epsilon\lambda}K_\epsilon \dot{z}_\epsilon + 2(\dot{P}_{\epsilon\lambda}\dot{z}_\epsilon)^i - 2(P_{\epsilon\lambda}\Gamma_\epsilon(\dot{y}_\epsilon, \dot{z}_\epsilon))^i \\ + a_{\epsilon\lambda jk}{}^i \dot{z}_\epsilon^j \dot{z}_\epsilon^k + \epsilon^{-2}b_{\epsilon\lambda jk}{}^i z_\epsilon^j z_\epsilon^k + c_{\epsilon\lambda}{}^i. \quad \text{(II.56)}$$

This additional term $-P_{\epsilon\lambda}K_\epsilon \dot{z}_\epsilon$ is the reason, that there is no adiabatic *invariance* in Lemma 14 any more. Instead, there holds the following exponential *fade-out* of the normal actions

$$\theta_\epsilon^\lambda \to \theta_0^\lambda = e^{-\kappa t}\theta_*^\lambda.$$

The rest of Step 3 provides a proof for this assertion. If we follow the proof of Lemma 14 with the evaluation of $\dot{E}_{\epsilon\lambda}^\perp$, there appears the additional term

$$s_a = -\operatorname{tr}(P_{\epsilon\lambda} K_\epsilon \dot{z}_\epsilon \otimes \dot{z}_{\epsilon\lambda} G_\epsilon) = -\sum_\mu \operatorname{tr}(P_{\epsilon\lambda} K_\epsilon \dot{z}_{\epsilon\mu} \otimes \dot{z}_{\epsilon\lambda} G_\epsilon)$$

$$= -\sum_\mu \operatorname{tr}(K_\epsilon P_{\epsilon\mu} \Pi_\epsilon P_{\epsilon\lambda}) + O(\epsilon).$$

By Eq. (II.37), we have the weak*-limit $s_a \overset{*}{\rightharpoonup} -\operatorname{tr}(K_0 P_{0\lambda} \Pi_0 P_{0\lambda})$. Now, the splitting $K = A + S$, with S being skew, yields

$$\operatorname{tr}(K_0 P_{0\lambda} \Pi_0 P_{0\lambda}) = \operatorname{tr}(A_0 P_{0\lambda} \Pi_0 P_{0\lambda}) = \kappa \operatorname{tr}(P_{0\lambda} \Pi_0 P_{0\lambda}) = \kappa \sigma_\lambda \omega_{0\lambda}^2.$$

Here, we have used the assumption that A is κ-isotropic, which allows to simplify $P_{0\lambda} A_0 P_{0\lambda} = \kappa \cdot P_{0\lambda}$ by Eq. (II.54). Summarizing, we obtain the weak*-limit

$$\dot{E}_{\epsilon\lambda}^\perp \overset{*}{\rightharpoonup} \dot{E}_{0\lambda}^\perp = -\kappa \sigma_\lambda \omega_{0\lambda}^2 + \tfrac{1}{2} \sigma_\lambda \frac{d}{dt} \omega_{0\lambda}^2.$$

A comparison with a direct differentiation of the expression $E_{0\lambda}^\perp = \sigma_\lambda \omega_{0\lambda}^2$, that is

$$\dot{E}_{0\lambda}^\perp = \dot{\sigma}_\lambda \omega_{0\lambda}^2 + \sigma_\lambda \frac{d}{dt} \omega_{0\lambda}^2,$$

yields the following equation of logarithmic differentials:

$$\frac{\dot{\sigma}_\lambda}{\sigma_\lambda} = -\kappa - \frac{\dot{\omega}_{0\lambda}}{\omega_{0\lambda}}.$$

Thus, there are constants θ_*^λ such that

$$\sigma_\lambda = \frac{e^{-\kappa t} \theta_*^\lambda}{\omega_{0\lambda}}, \qquad E_{0\lambda}^\perp = e^{-\kappa t} \theta_*^\lambda \omega_{0\lambda}, \qquad \theta_0^\lambda = e^{-\kappa t} \theta_*^\lambda.$$

As in the proof of Lemma 14, these constants can be calculated from the initial values by evaluating the limit of the energies $E_{\epsilon\lambda}^\perp$ at the initial time $t = 0$. To be specific, we obtain

$$E_{\epsilon\lambda}^\perp(0) = T_{\epsilon\lambda}^\perp(0) \;\to\; \tfrac{1}{2} \langle P_\lambda(x_*) v_*, P_\lambda(x_*) v_* \rangle,$$

which yields the values

$$\theta_*^\lambda = \theta_0^\lambda(0) = \lim_{\epsilon \to 0} \frac{E_{\epsilon\lambda}^\perp(0)}{\omega_\lambda(y_\epsilon(0))} = \frac{\langle P_\lambda(x_*) v_*, P_\lambda(x_*) v_* \rangle}{2 \omega_\lambda(x_*)},$$

just as in Definition 4.

Step 4: Identification of the Limit Mechanical System (as in §2.4). The assertion of Lemma 15 has to be modified as

$$Q(x_0) F_U^{\text{hom}} = e^{-\kappa t} \operatorname{grad}_N U_{\text{hom}}(x_0).$$

This can be proven as follows: literally as in the proof of Lemma 15, we first obtain

$$\tfrac{1}{2}\operatorname{tr}\left(\frac{\partial H(x_0)}{\partial y^i}\cdot \Sigma_0\right) = \sum_\lambda \sigma_\lambda \omega_\lambda(x_0)\cdot \frac{\partial \omega_\lambda(x_0)}{\partial y^i}.$$

Next, inserting the relation $\sigma_\lambda = e^{-\kappa t}\theta_*^\lambda/\omega_\lambda(x_0)$ of the previous step yields the desired result.   □

We can interpret the assertion of Theorem 4 depending on which value the constant κ of friction takes:

- $\kappa = 0$, the "purely" gyroscopic case normal to N. Here, we have $\mathscr{L}^\kappa_{\text{hom}} = \mathscr{L}_{\text{hom}}$ which means that Theorem 1 generalizes in perfect analogy.

- $\kappa > 0$, the case of friction normal to N. Here, the presence of the homogenized potential *fades out*, exponentially fast in time; yielding a continuous transition to the Lagrangian of holonomic constraints,

$$\mathscr{L}^\kappa_{\text{hom}}\big|_{t=0} = \mathscr{L}_{\text{hom}}, \qquad \lim_{t\to\infty}\mathscr{L}^\kappa_{\text{hom}} = \mathscr{L}_{\text{con}}.$$

This way, we have given a precise mathematical meaning to the physical argument of KOPPE and JENSEN [58, p. 8] that any strong potential *finally* realizes holonomic constraints if there is friction in the *normal* motion.[45]

As promised in Step 2 of the proof of Theorem 4, we will shortly discuss the splitting of energies that modifies Lemma 6. We will do so for all external forces of the form $F(x,\dot{x}) = -K(x)\dot{x}$, K being nonnegative; i.e.,

---

[45]KOPPE and JENSEN write (loc. cit.): "Da aber die ganze Technische Mechanik durch das d'Alembertsche Prinzip beherrscht wird und es wohl nur wenige besser empirisch bestätigte Naturgesetze gibt, ist notwenigerweise außer der Annahme eines Führungspotentials (bzw. von Führungskräften) zu seiner Begründung noch ein weiterer Gesichtspunkt erforderlich. Dieser scheint uns darin zu liegen, daß die [...] ‚Energie der Transversalbewegung' wegen der hohen Frequenz, mit der in ihr sich kinetische und potentielle Energie ineinander umsetzen, sehr rasch dissipiert wird, auch wenn wir von der Dämpfung der Bewegung längs der Führungsgeraden noch ganz absehen können. Infolgedessen wird, wie auch immer die Bewegung gestartet sein mag, [die transversale Energie] durch Dämpfung rasch gegen Null gehen, und dann die weitere Bewegung so verlaufen, wie sie durch das d'Alembertsche Prinzip bestimmt ist."

[Translation by the author: "Because all of the mechanics in technical applications is governed by d'Alembert's principle, and hardly a law of nature is empirically better confirmed, there is necessarily a further aspect required in addition to the assumption of a constraining potential (resp. a constraining force). In our view, this aspect seemingly lies in the fact that the 'energy of transversal motion' is rapidly dissipated—even if we may completely neglect all damping of the motion along the constraint manifold—because kinetic energy is converted to potential energy and vice versa with high frequency. Therefore, however the motion is started, the transversal energy will rapidly be damped to zero and then, afterwards, the motion will take place according to d'Alembert's principle."]

we drop the assumption of κ-isotropy of the friction from now on. The reader should notice, that we have not used κ-isotropy during the first two steps of the proof of Theorem 4.

Leaving the notion of the energy $E_\epsilon^\perp = T_\epsilon^\perp + U_\epsilon^\perp$ of the normal motion as in Definition 5, we modify the energy of the constrained motion to

$$E_\epsilon^\| = \tfrac{1}{2}\langle \dot{y}_\epsilon, \dot{y}_\epsilon \rangle + V(y_\epsilon) + \int_0^t \langle K(y_\epsilon)\dot{y}_\epsilon, \dot{y}_\epsilon \rangle \, d\tau$$

and additionally define the energy that is *dissipated* in normal direction,

$$E_\epsilon^{\mathrm{diss}} = \int_0^t \mathrm{tr}(K_\epsilon(\tau) \cdot \Pi_\epsilon(\tau)) \, d\tau \;\rightarrow\; E_0^{\mathrm{diss}} = \int_0^t \mathrm{tr}(K(x_0) \cdot \Pi_0) \, d\tau.$$

By slightly adjusting the proof of Lemma 6, one obtains the following result: As $\epsilon \to 0$, the total energy E_ϵ decomposes into $E_\epsilon = E_\epsilon^\| + E_\epsilon^\perp + E_\epsilon^{\mathrm{diss}} + o(1)$. All three components converge uniformly as functions in $C[0,T]$.

Now, using the abstract limit equation (II.55) and arguing analogously to the proof of Eq. (II.48), we obtain

$$\begin{aligned} E_0^\perp(t) &= -E_0^{\mathrm{diss}}(t) + E_0^\perp(0) + \tfrac{1}{2}\int_0^t \langle \dot{x}_0(\tau), \mathrm{grad}\, H(x_0(\tau)) : \Sigma_0(\tau) \rangle \, d\tau \\ &\leq E_0^\perp(0) + \tfrac{1}{2}\int_0^t \langle \dot{x}_0(\tau), \mathrm{grad}\, H(x_0(\tau)) : \Sigma_0(\tau) \rangle \, d\tau. \end{aligned}$$

$$(\text{II.57})$$

For the last estimate, we notice that the nonnegativity of K implies that $E_\epsilon^{\mathrm{diss}} \geq 0$ which in turn yields $E_0^{\mathrm{diss}} \geq 0$.

This estimate allows to prove results on realization of holonomic constraints that are completely analogous to §3—extending the Definition 8 of the notion "realization of holonomic constraints" to the presence of external forces $F(x, \dot{x}) = -K(x)\dot{x}$, K being nonnegative, in the obvious way. Using estimate (II.57) instead of Eq. (II.48) in the proof of Theorem 2, and the abstract limit equation (II.55) instead of Eq. (II.30) in the proof of Theorem 3, we obtain the following theorem.

Theorem 5.   (i) *A pair (x_*, v_*) of initial values, $x_* \in N$ and $v_* \in T_{x_*}M$, realizes holonomic constraints for all potentials U constraining to N and all external forces $F(x, \dot{x}) = -K(x)\dot{x}$ (K nonnegative), if and only if $v_* \in T_{x_*}N$.*

   (ii) *Let U be constraining to the critical manifold N. Suppose the spectrum $\sigma(H)$ of the Hessian H is constant on N. Then, the potential U realizes holonomic constraints for all external forces $F(x, \dot{x}) = -K(x)\dot{x}$ (K nonnegative) and for all initial values $x_* \in N$, $v_* \in T_{x_*}M$.*

Notice that no resonance condition applies.

Finally, the characterization of strong convergence given in Lemma 17 remains true, with literally the same proof.

III

Applications

Suppose that, due to strong forcing, the motion of a mechanical system exhibits rapid oscillations around some equilibrium positions. These equilibrium positions form a submanifold N of the configuration space M. If the rapid oscillations occur on a time scale τ_{fast}, compared to the time scale τ_{avg} of the *average* motion, they introduce a small scale ratio

$$\epsilon = \frac{\tau_{\text{fast}}}{\tau_{\text{avg}}} \ll 1.$$

For a variety of reasons, one might be interested to establish models which approximate the average motion of the mechanical system by a dynamical system on N, thus *eliminating* the rapidly oscillating degrees of freedom. These reasons are, e.g.:

- The understanding of the average motion, cf. §§1 and 3.

- A simplification of the model, and a dimensional reduction from the configuration space M to the submanifold N, cf. §§1 and 2.

- The acceleration of numerical integrators which suffer from small time-scales which require the use of correspondingly small time-steps, cf. §§2 and 3.

The homogenization theory of Chapter II offers, if applicable, a zero-order approximation in ϵ, i.e., the singular limit $\epsilon \to 0$. As far as the present author knows, there are *no* results available which provide higher-order approximations for submanifolds N of codimension *higher* than one.[46] Even the formal multiscale asymptotics of KELLER and RUBINSTEIN [52] provides just the zero order approximation.

If the mechanical system is not given in a form that Theorem II.1 is directly applicable, one nevertheless might be able to *transform* it into a natural mechanical system with a strong constraining potential. Two important transforms should be considered:

---

[46]With the exception, however, of the two exemplary asymptotic studies of Appendix C, added later by the present author. In fact, those studies show how involved an asymptotic analysis of higher order would be in general.

- Dimensional reduction of systems with symmetry. An example is provided by the method of Routh for eliminating cyclic variables, cf. §1.

- Legendre transform of a system given in Hamiltonian formulation, cf. §3.

In the present chapter, we will discuss three applications from different areas of the natural sciences. The first one, in §1, is from *plasma physics*. The second one, in §2, is from (classical) *molecular dynamics*. The third one, in §3, is from *quantum chemistry*. This final application is the starting point of Chapter IV, where we will extend the methods of Chapter II to an infinite-dimensional problem.

§1.   Magnetic Traps and Mirrors

Here, we discuss a classical, very interesting application of the homogenization results in the codimension $r = 1$ case. We will closely follow the discussion given by BORNEMANN and SCHÜTTE [18, §6.2].

We consider the motion of a nonrelativistic, charged particle in a nonuniform, steady magnetic field $B(x)$. We assume that this field is "strongly axially symmetric" in the sense that, in cylindrical coordinates r, z, ϕ, the B-field does not depend on the angle ϕ and also the ϕ-component of B vanishes. We express the divergence-free field B in terms of a vector potential $\mathscr{A}$ with components $\mathscr{A} = (0, 0, A(r, z))$, such that $B = \operatorname{curl} \mathscr{A}$, i.e.,

$$B = (-\partial_z A, \partial_r A + A/r, 0). \tag{III.1}$$

We denote by e the charge of the particle and by m its mass.

According to [61, §17], the Lagrangian of the motion is given by

$$\mathscr{L} = \tfrac{1}{2} m \langle \dot{x}, \dot{x} \rangle + e \langle \dot{x}, \mathscr{A} \rangle = \tfrac{1}{2} m (\dot{r}^2 + \dot{z}^2 + r^2 \dot{\phi}^2) + e r \dot{\phi} A.$$

Here, $x = (x^1, x^2, x^3)$ and $\langle \cdot, \cdot \rangle$ denotes the Euclidean inner product of $\mathbb{R}^3$. Since $\mathscr{L}$ does not depend on ϕ, we obtain the conservation of the angular momentum,

$$\frac{\partial \mathscr{L}}{\partial \dot{\phi}} = J = \text{const, i.e.,} \qquad m r^2 \dot{\phi} + e r A = J.$$

By the classical method of Routh [6, Chap. 3, §2.1], we eliminate the cyclic variable ϕ reducing the Lagrangian in (r, z)-coordinates to

$$\mathscr{L}_{\text{red}} = \mathscr{L} - J\dot{\phi} \Big|_{m r^2 \dot{\phi} + e r A = J} = \tfrac{1}{2} m (\dot{r}^2 + \dot{z}^2) - \tfrac{1}{2} m^{-1} e^2 (A - J/er)^2.$$

Since multiplication of the Lagrangian by the constant factor m^{-1} does not change the equation of motion, we instead consider the Lagrangian

$$\mathscr{L}_\epsilon = \tfrac{1}{2} (\dot{r}^2 + \dot{z}^2) - \epsilon^{-2} U(r, z), \qquad U(r, z) = \tfrac{1}{2} (A - J/er)^2, \qquad \epsilon = m/e.$$

This fits into the framework of Chapter II. The results of §§II.1-2 show that, for a large *specific charge* $\epsilon^{-1} = e/m$, the projection of the motion to the (r, z)-plane oscillates very rapidly in a small neighborhood of the line

$$N_{\text{red}} = \{(r, z) : A(r, z) = J/er\}.$$

In addition, we are able to describe the secular oscillations of the angular variable ϕ. We have to use some of the notions and results of §II.2.1. In a vicinity of N_{red} we introduce tubular coordinates $(\zeta^{\parallel}, \zeta^{\perp})$ of the (r, z)-plane. This means that $\zeta^{\parallel}$ denotes the nearest point on N_{red} and $\zeta^{\perp}$ its distance to the point under consideration. By Lemma 4 of §II.2.1, we obtain[47]

$$\dot{\phi}_{\epsilon} = \frac{\epsilon^{-1}}{r_{\epsilon}} \left(\frac{J}{er_{\epsilon}} - A(r_{\epsilon}, z_{\epsilon}) \right) = \pm\sqrt{2\epsilon^{-2}U}/r_{\epsilon}$$

$$= \frac{\omega(\zeta_{\epsilon}^{\parallel})}{r_{\epsilon}} \epsilon^{-1}\zeta_{\epsilon}^{\perp} + O(\epsilon^{1/2}) \xrightarrow{\ast} 0,$$

implying by the extended Arzelà-Ascoli theorem, Principle I.4, the uniform convergence $\phi_{\epsilon} \to \phi_{\ast}$ to the fixed initial value $\phi_{\ast} = \phi_{\epsilon}(0)$. Thus, the actual motion in *space* is a small amplitude *gyration* around the line

$$N = \{(r, z, \phi) : A(r, z) = J/er, \phi = \phi_{\ast}\},$$

the so-called *guiding center* of the motion which, in fact, is a field line of the magnetic field (III.1).

The frequency of gyration is $\epsilon^{-1}\omega$, where ω^2 is given as the single nonzero eigenvalue of the Hessian

$$H = D^2U|_{N_{\text{red}}} = \left. \begin{pmatrix} (\partial_r A + J/er^2)^2 & \partial_z A \cdot (\partial_r A + J/er^2) \\ \partial_z A \cdot (\partial_r A + J/er^2) & (\partial_z A)^2 \end{pmatrix} \right|_{N_{\text{red}}}$$

$$= \begin{pmatrix} B_z^2 & -B_r B_z \\ -B_r B_z & B_r^2 \end{pmatrix},$$

computed in (r, z)-coordinates. Here, B_r and B_z denote the components of the B-field in the corresponding directions. A simple calculation reveals

$$\omega^2 = B_r^2 + B_z^2 = |B|^2.$$

Alternatively, one can use the result of Example II.1. With

$$\text{grad}\,(A - J/er)|_{N_{\text{red}}} = (B_z, -B_r),$$

---

[47]Notice that the component $\zeta_{\epsilon}^{\perp}$ is just z_{ϵ} of Lemma II.4, cf. Eq. (II.21).

using (r, z)-coordinates, one likewise obtains $\omega = |B|$. Thus, just as in the case of a uniform magnetic field, the particle gyrates with the *Larmor frequency* $e|B|/m$.

Theorem 1 shows that in the limit $\epsilon \to 0$ the average tangential motion along the guiding center N is governed by the potential

$$U_{\mathrm{hom}} = \theta_* |B|,$$

where the *adiabatic invariant* θ_* constitutes the *magnetic moment* of the particle motion. Therefore, the limit equation of motion now reads

$$\ddot{s} = -\theta_* \frac{\partial}{\partial s} |B|, \tag{III.2}$$

where s denotes arc length on the line N.

As we see, the appearing homogenized potential U_{hom} introduces the *only* force term for the limit motion. This force term is of utmost importance in engineering and natural sciences: Charged particles are moderated by an increasingly strong magnetic field—and that the more, the bigger the initial normal velocity was. This is the working principle of *magnetic traps*[48] and *magnetic mirrors* in plasma physics, as well as of the Van Allen radiation belt of the earth with all its implications for northern lights and astronautics.

Remark 1. The first derivation of equation (III.2) by physical reasoning was given by the Swedish Nobel prize winner ALFVÉN [2, Chapter 2.3], see also the books of NORTHROP [72] and SPITZER [93].

The guiding center motion (III.2) was derived by HELLWIG [44] and KRUSKAL [60] by means of a formal two-scale, or WKB-type, asymptotics. Later on, BERKOWITZ and GARDNER [15] derived error bounds for these formal asymptotic expansions.

The first mathematically complete discussion of the limit $e/m \to \infty$ was given by RUBIN and UNGAR [82] who also discuss a nice mechanical analogue of the magnetic mirror. However, they only considered the *reduced* motion in the (r, z)-plane. The adiabatic invariance of the magnetic moment was shown by ARNOLD in his seminal paper [3] using now well established tools[49] of the perturbation theory of integrable Hamiltonian systems.

§2. Molecular Dynamics

In classical molecular dynamics approaches, the simulation of the dynamical behavior of a molecular system is based on the assumption that the system of interest is well predictable using classical mechanics. The molecule

---

[48] For obvious reasons, they are sometimes called *adiabatic traps*.
[49] Only slightly more elaborate than the discussion of codimension $r = 1$ in §I.2.6.

is viewed as being composed of m mass points with positions $x_i \in \mathbb{R}^3$. Using a more or less empirically constructed interaction potential W, the dynamics is assumed to be governed by the Lagrangian

$$\mathscr{L}(x, \dot{x}) = \tfrac{1}{2}\langle G\dot{x}, \dot{x}\rangle - W(x), \qquad x \in \mathbb{R}^{3m}, \ \dot{x} \in \mathbb{R}^{3m}.$$

Here, $G \in \mathbb{R}^{3m \times 3m}$ denotes the diagonal *mass matrix* and $\langle \cdot, \cdot \rangle$ the Euclidean inner product of $\mathbb{R}^{3m}$. Typically, the potential can be split into two parts of essentially different strength. Indicating this, we write the interaction potential as the sum

$$W(x) = V_{\text{weak}}(x) + V_{\text{strong}}(x), \qquad V_{\text{strong}}(x) = \epsilon^{-2}U(x),$$

where V_{strong} represents the strong interactions and V_{weak} the collection of all weak interactions. The scaling of U and ϵ is chosen in the following way: Let τ_{fast} be a period of a typical fast vibration for which V_{strong} is responsible, and τ_{ref} a reference time-scale of the average motion. We then define

$$\epsilon = \frac{\tau_{\text{fast}}}{\tau_{\text{avg}}}.$$

There is a strong practical need for eliminating these small time scales because they pose a severe restriction for efficient and reliable long-term numerical simulations of macromolecules.

The potential V_{strong} usually collects the bond-stretching and bond-angle potentials which have unique equilibria. Thus, we can assume that U is constraining to the manifold N of fixed, or "frozen", bond-lengths and bond-angles. If the strong potential U happens to constrain *spectrally smooth*,[50] the homogenization theory of Chapter II is applicable. Theorem II.1 teaches that a zero-order description of the motion is given by the Lagrangian

$$\mathscr{L}_{\text{hom}}(x, \dot{x}) = \tfrac{1}{2}\langle G\dot{x}, \dot{x}\rangle - V_{\text{weak}}(x) - U_{\text{hom}}(x), \qquad \dot{x} \in T_x N.$$

According to Definition II.4, the homogenized potential is given by the expression

$$U_{\text{hom}}(x) = \sum_{\lambda} \theta_*^{\lambda}\, \hat{\omega}_{\lambda}(x), \qquad x \in N.$$

---

[50]We have no idea, whether in general the strong potentials of molecular dynamics are *generically* constraining spectrally smooth. By the arguments of §II.1.7, this is certainly the case as long as there are no resonances of the normal frequencies. Otherwise, it will depend on the admissible type of perturbations that define the term "generic." Now, because the strong potentials are usually of short range, typically nearest neighbor, the Hessian will have a block diagonal structure. Only perturbations that preserve this structure are "physical." Due to this restriction it can happen that a resonance is of codimension one only, or if of higher codimension *non-generic* because the parameter-dependence of the corresponding eigenspace is essentially only one-dimensional. This latter effect occurs in the Example of §2.1, the *butane molecule*, which constitutes a typical building block of the block diagonal structure previously mentioned.

Here, the *frequencies* $\hat{\omega}_\lambda$ are given via the spectral decomposition of the Hessian H of U,

$$H(x) = G^{-1}D^2U(x) = \sum_\lambda \hat{\omega}_\lambda^2(x)\, P_\lambda(x), \qquad x \in N.$$

Using Definition II.7 and Lemma II.14, the adiabatic invariants θ_*^λ can be expressed as the ratios

$$\theta_*^\lambda = \frac{E_\lambda^\perp(0)}{\hat{\omega}_\lambda(x_*)},$$

where $E_\lambda^\perp(0)$ denotes the energy in the λ-vibrational mode at time $t = 0$, and the point x_* the orthogonal projection of the initial position to the constraints manifold N.

The reader should notice that the homogenized potential, U_{hom}, does *not* depend on the actual scaling of U and ϵ. Denoting by ω_λ the frequencies of the strong potential, V_{strong}, itself,

$$\omega_\lambda = \epsilon^{-1}\hat{\omega}_\lambda,$$

we obtain the scaling invariance

$$U_{\text{hom}}(x) = \sum_\lambda E_\lambda^\perp(0)\frac{\hat{\omega}_\lambda(x)}{\hat{\omega}_\lambda(x_*)} = \sum_\lambda E_\lambda^\perp(0)\frac{\omega_\lambda(x)}{\omega_\lambda(x_*)} = (V_{\text{strong}})_{\text{hom}}(x).$$

This is, because energy is not effected by the scaling.

However, establishing a proper scaling of the singular perturbation parameter ϵ can help to understand the capabilities of the approach. The smaller the size of ϵ is, the better will be the approximation of the zero-order model and, equally important, the greater will be the pay-off in computational work: As a rule of thump, eliminating a time scale of the order $O(\epsilon)$ will increase the time-step of a numerical integrator by a factor of $O(\epsilon^{-1})$.

§2.1.   An Example: The Butane Molecule

We illustrate the possibilities of the approach with one of the simplest *realistic* examples, the lumped butane molecule. We follow closely the presentation previously given by SCHÜTTE and BORNEMANN [89]. The data of the butane molecule are taken from [99].

The model of the butane molecule, CH_3-CH_2-CH_2-CH_3, consists of four mass points given by the four CH-groups. These groups are located at the positions $x_i \in \mathbb{R}^3$, $i = 1, \ldots, 4$. Thus, the configuration space is the 12-dimensional space $\mathbb{R}^{12}$. The mass matrix is given by

$$G = \text{diag}(m_1, m_1, m_1, m_2, m_2, m_2, m_2, m_2, m_2, m_1, m_1, m_1),$$

according to the mass $m_1 = 15u$ of the CH_3-group, and $m_2 = 14u$ of the CH_2-group. Here, "u" denotes the atomic mass unit, $1u = 1.66054 \cdot 10^{-27}$kg. The strong part of the interaction potential W is given by the bond-stretching and bond-angle potentials,

$$V_{\text{strong}}(x) = \sum_{k=1}^{3} V_{\text{bs}}(x_k, x_{k+1}) + V_{\text{ba}}(x_1, x_2, x_3) + V_{\text{ba}}(x_2, x_3, x_4). \quad \text{(III.3)}$$

The bond-stretching potential V_{bs} is modeled as a spring,

$$V_{\text{bs}}(x_k, x_{k+1}) = \tfrac{1}{2}\alpha(|x_k - x_{k+1}| - r_0)^2,$$

where $|\cdot|$ denotes Euclidean length in $\mathbb{R}^3$. The equilibrium length r_0 and the constant α are given by

$$r_0 = 1.53\,\text{Å}, \qquad \alpha = 83.7\,\frac{\text{kcal}}{\text{mol} \cdot \text{Å}^2}.$$

As usual in chemistry, we have $1\text{kcal} = 4184\text{J}$. The bond-angle potentials V_{ba} are "quasi-harmonically" given as

$$V_{\text{ba}}(x_k, x_{k+1}, x_{k+2}) = \tfrac{1}{2}\beta(\cos\phi - \cos\phi_0)^2.$$

Here, $\phi = \phi(x_k, x_{k+1}, x_{k+2})$ denotes the angle between the bonds connecting x_k with x_{k+1}, and x_{k+1} with x_{k+2},

$$\cos\phi(x_k, x_{k+1}, x_{k+2}) = \frac{(x_k - x_{k+1})^T (x_{k+2} - x_{k+1})}{|x_k - x_{k+1}||x_{k+2} - x_{k+1}|}.$$

The equilibrium angle ϕ_0 and the constant β are given by

$$\phi_0 = 109.5°, \qquad \beta = 43.1\,\frac{\text{kcal}}{\text{mol}}.$$

The weak part V_{weak} of the interaction potential W is given by the so-called *torsion-angle potential*

$$V_{\text{weak}}(x) = V_{\text{tor}}(\cos\theta(x)),$$

with

$$V_{\text{tor}}(c) = (1.116 - 1.462c - 1.578c^2 + 0.368c^3 + 3.156c^4 + 3.788c^5)\,\frac{\text{kcal}}{\text{mol}}.$$

The torsion-angle $\theta = \theta(x_1, x_2, x_3, x_4)$ is the angle between the two planes spanned by x_1, x_2, x_3 and x_2, x_3, x_4, respectively,

$$\cos\theta(x) = \frac{(r_1 \times r_2)^T (r_2 \times r_3)}{|r_1 \times r_2||r_2 \times r_3|}, \qquad r_k = x_{k+1} - x_k.$$

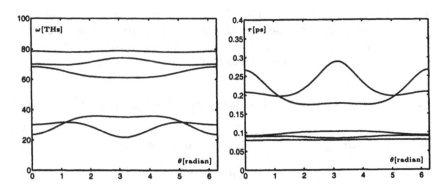

Figure III.1: The normal frequencies ω (left) and corresponding periods τ (right).

The torsion-angle potential $V_{tor}(\cos\theta)$ is symmetric with respect to $\theta = \pi$, cf. Figure III.2, and possesses a primary minimum located at $\theta = \pi$, and two secondary minima near $\theta = \pi/3$ and $\theta = 5\pi/3$. These minima account for the two different geometrical *conformations* of the butane molecule.

The strong potential (III.3) is of the form discussed in §II.1.6. It is a sum of five quadratic terms, therefore we have $r = 5$. We number the five terms of the right hand side of Eq. (III.3) according to their appearance from left to right. For computing the five frequencies ω_k, $k = 1, \ldots, 5$, we first establish the reduced Hessian H_r as defined in Eq. (II.17). A lengthy calculation, which we omit here, reveals that

$$H_r(x) = \begin{pmatrix} \eta_1 & \eta_2 & 0 & \eta_4 & \eta_5 \\ \eta_2 & \eta_3 & \eta_2 & \eta_4 & \eta_4 \\ 0 & \eta_2 & \eta_1 & \eta_5 & \eta_4 \\ \eta_4 & \eta_4 & \eta_5 & \eta_7 & \eta_6 \\ \eta_5 & \eta_4 & \eta_4 & \eta_6 & \eta_7 \end{pmatrix}, \qquad x \in N,$$

where we use the abbreviations

$$\eta_1 = \frac{\alpha}{m_1} + \frac{\alpha}{m_2}, \quad \eta_2 = \frac{\alpha \cos\phi_0}{m_2}, \quad \eta_3 = \frac{2\alpha}{m_2}, \quad \eta_4 = -\frac{\sqrt{\alpha\beta}\,(1 + \cos\phi_0)^2}{r_0\,m_2},$$

$$\eta_5 = -\frac{\sqrt{\alpha\beta}\,\cos\theta\,\sin^2\phi_0}{r_0\,m_2}, \quad \eta_6 = \frac{2\beta\,(1 + \cos\phi_0)\cos\theta\,\sin\phi_0^2}{r_0^2\,m_2},$$

$$\eta_7 = \frac{\beta\,(1 + \cos\phi_0)}{r_0^2} \left(\left(\frac{1}{m_1} + \frac{1}{m_2}\right)(1 - \cos\phi_0) + \frac{2}{m_2}(1 + \cos\phi_0)^2 \right).$$

Notice that the reduced Hessian H_r does only depend on the torsion angle θ. This single-parameter dependence can be explained as follows: Due to conservation of momentum and angular momentum, the configuration

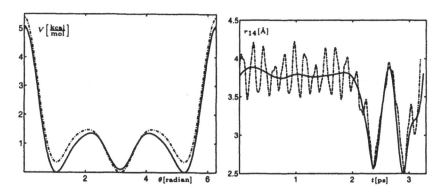

Figure III.2: Left: Torsion angle potential V_{weak} (dashed line) and the corrected potential $V_{\text{weak}} + (V_{\text{strong}})_{\text{hom}}$ (solid line). Both potentials are shifted such that their global minimum has value zero. Right: Evolution of the distance $|x_1 - x_4|$, for the original solution (dashed line) and for the homogenized solution (solid line).

space is 6-dimensional effectively. The intersection of the constraint manifold N with this effective configuration space still has codimension 5, i.e., it is one-dimensional.

By a standard result of perturbation theory [51, Chap. 2, Thm. 6.1], this *holomorphic* dependence of H_r on the *single* parameter θ yields a smooth spectral decomposition of H_r. Moreover, the smoothness of the spectral decomposition is *generic*, if only perturbations of the parameters α, β, r_0, ϕ_0, m_1 and m_2 are allowed being considered as physically reasonable perturbations of the potential.

Computed from the eigenvalue decomposition of H_r, the five frequencies $\omega_1, \ldots, \omega_5$ are shown as functions of the torsion angle θ in Figure III.1. We number them according to

$$\omega_1(\pi) < \omega_2(\pi) < \omega_3(\pi) < \omega_4(\pi) < \omega_5(\pi).$$

Because of

$$\omega_j(\theta) \geq 20\,\text{THz}, \qquad \theta \in [0, 2\pi], \quad j = 1, \ldots, 5,$$

the nondegeneracy condition (II.10) is fulfilled.

Summarizing, the strong potential V_{strong} is generically constraining spectrally smooth to the equilibrium manifold of "frozen" bond-lengths and bond-angles,

$$\begin{aligned} N \;=\; &\{x \in \mathbb{R}^{12} : |x_2 - x_1| = |x_3 - x_2| = |x_4 - x_3| = r_0, \\ &\phi(x_1, x_2, x_3) = \phi(x_2, x_3, x_4) = \phi_0\}\end{aligned}$$

of codimension $r = 5$. We notice that, in an obvious way, the frequencies ω_1 and ω_2 as well as ω_3, ω_4, and ω_5 form two groups. The first group of lower frequencies corresponds to the two bond-angle potentials, and the second of higher frequencies to the three bond-stretching potentials. This reflects the experience that the bond-stretching potentials are "stronger" than the bond-angle potentials. We further notice that there are two *resonance* points of the frequencies ω_1 and ω_2. These resonances occur at torsion angles which are near to the secondary minima of the weak potential, cf. Figure III.2. Corresponding to the frequency ω_j, there is a "period"

$$\tau_j = \frac{2\pi}{\omega_j}, \qquad j = 1, \ldots, 5.$$

These periods are also shown in Figure III.1. The "fast" time scale of V_{strong}, as compared to V_{weak}, is given by the maximum value of the τ_j, i.e.,

$$\tau_{\text{fast}} \approx 0.25 \, \text{ps}.$$

As it turns out in numerical simulations, cf. Figure III.2, a typical average time scale would be given by $\tau_{\text{avg}} = 1 \, \text{ps}$ yielding a scale ratio

$$\epsilon \approx 0.25,$$

which cannot really be considered as small.

For comparing the original motion with the zero order approximation of the homogenization $\mathscr{L}_{\text{hom}}$, we choose initial data for which

$$E_4^\perp(0) = 3.577 \, \frac{\text{kcal}}{\text{mol}}, \qquad E_1^\perp(0) = E_2^\perp(0) = E_3^\perp(0) = E_5^\perp(0) = 0.$$

This energy amounts for half of the average energy of the butane molecule in a gas at temperature 300K. The left subfigure of Figure III.2 shows how the homogenized potential $(V_{\text{strong}})_{\text{hom}}$ "corrects" (solid line) the torsion potential V_{weak} (dashed line). In particular, the roles of the primary and secondary minima are reversed. The right subfigure of Figure III.2 shows the "length" of the molecule, $r_{14} = |x_4 - x_1|$, for the original motion (dashed) and the homogenized motion (solid line). We observe a good approximation of the mean value despite the fact that the given scale ratio is rather large. The time-step of the numerical integrator increases by a factor which is roughly the scale ratio of the smallest eliminated time scale,

$$\frac{\tau_{\text{avg}}}{\tau_5} \approx 10.$$

For larger times, the homogenized solution increasingly deviates from the original motion. Given a fixed $\epsilon > 0$, this had to be expected . However, the good approximation for intermediate times is promising in two respects:

- We gained analytical insight into the model of the butane molecule. In particular, one might consider to *remodel* according to the homogenization with appropriately chosen normal energies $E_j^\perp(0)$.

- The understanding of the singular limit $\epsilon \to 0$ opens the way towards advanced numerical techniques for simulating the originally given model efficiently. In our example, the increase of the time step by an order of magnitude points in this direction.

§2.2. Relation to the Fixman-Potential

Historically, people in the molecular dynamics business first tried to establish a zero-order model with frozen bond-lengths and bond-angles using just the Lagrangian of holonomic constraints

$$\mathscr{L}_{\mathrm{con}}(x, \dot{x}) = \tfrac{1}{2}\langle G\dot{x}, \dot{x}\rangle - V_{\mathrm{weak}}(x), \qquad \dot{x} \in T_x N.$$

It soon became clear that the results are unsatisfactory. REICH [79] developed the idea that the model of holonomic constraints should be corrected by an additional potential. Using arguments of equi-distribution[51] of energy among the fast vibrational modes, borrowed from thermodynamics, he came up with the so called *Fixman-potential* of statistical physics, namely

$$U_{\mathrm{Fix}}(x) = \tfrac{1}{2}E^{\perp} \cdot \log \det H_r(x), \qquad x \in N,$$

where H_r denotes the *reduced* Hessian of U as introduced in §II.1.6. The constant $E^{\perp}$ stands for the energy in any of the normal vibrational modes. Thus, he suggested to use

$$\mathscr{L}_{\mathrm{Fix}}(x, \dot{x}) = \tfrac{1}{2}\langle G\dot{x}, \dot{x}\rangle - V_{\mathrm{weak}}(x) - U_{\mathrm{Fix}}(x), \qquad \dot{x} \in T_x N.$$

For short time-simulations, this potential correction showed quite promising results. This can be understood by the following argument: Equidistribution of energy in the normal vibrational modes yields that

$$E_{\lambda}^{\perp}(0) = n_{\lambda} \cdot E^{\perp}, \qquad \lambda = 1, \ldots, s.$$

Hence, the homogenized potential is given by

$$U_{\mathrm{hom}}(x) = E^{\perp} \sum_{\lambda} n_{\lambda} \frac{\omega_{\lambda}(x)}{\omega_{\lambda}(x_*)}, \qquad x \in N.$$

For $x \approx x_*$, such as for small times, the corresponding force term can by approximated by the force term of the Fixman-potential,

$$
\begin{aligned}
\operatorname{grad} U_{\mathrm{hom}}(x) &= E^{\perp} \sum_{\lambda} \frac{n_{\lambda}}{\omega_{\lambda}(x_*)} \operatorname{grad} \omega_{\lambda}(x) \\
&\approx E^{\perp} \sum_{\lambda} \frac{n_{\lambda}}{\omega_{\lambda}(x)} \operatorname{grad} \omega_{\lambda}(x) = \operatorname{grad} U_{\mathrm{Fix}}(x).
\end{aligned}
$$

---

[51]This should not be mixed up with the *provable* equi-partitioning of energy into its potential and kinetic part within each normal vibrational mode, i.e., Lemma II.13.

Here, we have used the relation

$$U_{\text{Fix}}(x) = \tfrac{1}{2}E^{\perp} \cdot \log\left(\prod_{\lambda} \omega_{\lambda}(x)^{2n_{\lambda}}\right) = E^{\perp}\sum_{\lambda} n_{\lambda}\log\omega_{\lambda}, \qquad x \in N.$$

However, the two forces are obviously not the same and the Fixman-potential does *not* describe the zero-order motion.

Remark 2. Interestingly enough, GALLAVOTTI [34, p. 172ff], in his discussion of the homogenization problem for the codimension one case, Example II.1, did not derive the correct homogenized potential U_{hom} but rather the Fixman-potential U_{Fix} instead. Studying the special case $\omega^2(x) = 1 + x^2$, however, "only in a heuristic, nonrigorous way", as he writes, he arrives at the potential

$$U_{\text{Fix}}(x) = \tfrac{1}{2}E^{\perp} \cdot \log(1 + x^2).$$

instead of

$$U_{\text{hom}}(x) = E^{\perp}\sqrt{\frac{1 + x^2}{1 + x_*^2}}$$

The reason for this flaw is exactly as above: He first *correctly* derives for $x \approx x_*$

$$\text{grad}\, U_{\text{hom}}(x) \approx E^{\perp}\frac{x}{1 + x_*^2},$$

but in turn he argues that $\text{grad}\, U_{\text{hom}}$ is therefore given by

$$E^{\perp}\frac{x}{1 + x^2},$$

which is not correct but yields U_{Fix}.

§3. Quantum-Classical Coupling: The Finite Dimensional Case

Quite recently, the coupling of quantum mechanical systems with systems behaving more classical has become an important modeling issue in molecular dynamics and quantum chemistry, cf., e.g., [9][10][14][35]. The model suggested in the latter two reference can be based on *first principles* as was shown by BORNEMANN, NETTESHEIM, and SCHÜTTE [17]. This model consists of a singularly perturbed Schrödinger equation that is nonlinearly coupled to a classical Newtonian equation of motion. For a further justification of the model, and for the development of efficient numerical methods, an understanding of the singular limit is extremely important. This singular limit and further mathematical aspects of the model are subject of Chapter IV.

Here, we consider a finite dimensional analogue—as for instance arising after a Galerkin approximation of the infinite dimensional Hamiltonian. We will show that the singular limit can be discussed by the homogenization theory that we have developed in Chapter II. This helps to understand, why only technical modifications of the basic method are required to handle the general case of Chapter IV.

The model system is given by the equations

$$
\begin{aligned}
\ddot{y}^j_\epsilon &= -\langle \partial_j H(y_\epsilon)\psi_\epsilon, \psi_\epsilon \rangle, \qquad j = 1, \ldots, n, \\
i\epsilon \dot{\psi}_\epsilon &= H(y_\epsilon)\psi_\epsilon.
\end{aligned}
\tag{III.4}
$$

Here, we denote by $y \in \mathbb{R}^n$ the "classical" positions, by $\psi \in \mathbb{C}^r$ the "quantum" states, and by $\langle \cdot, \cdot \rangle$ the Euclidean inner product[52] on $\mathbb{C}^r$. The *smooth* mapping $H : \mathbb{R}^n \to \mathbb{R}^{r \times r}$ takes *real symmetric matrices* as values, uniformly positive definite, i.e., for some $\omega_* > 0$ we have

$$
\langle H(y)\psi, \psi \rangle \geq \omega_* |\psi|^2, \qquad y \in \mathbb{R}^n, \ \psi \in \mathbb{C}^r.
\tag{III.5}
$$

The Euclidean norm on $\mathbb{R}^n$ and $\mathbb{C}^r$ will be denoted by $|\cdot|$. Finally, we specify initial values for (III.4) by

$$
y_\epsilon(0) = y_*, \qquad \dot{y}_\epsilon(0) = w_*, \qquad \psi_\epsilon(0) = \psi_*.
\tag{III.6}
$$

We assume,[53] that ψ_* has a *single* complex phase, i.e., there is a $\gamma \in \mathbb{C}$ with $|\gamma| = 1$ such that

$$
i\gamma\psi_* \in \mathbb{R}^r.
\tag{III.7}
$$

We will show that the coupling equations (III.4) are the equations of motion of a certain natural mechanical system to which Theorem II.1 is applicable. This will be accomplished in three steps. First, by making the equations real. Second, by giving them a canonical structure. Third, by applying a Legendre transform.

We introduce phase space coordinates $(y, \eta; z, \zeta) \in \mathbb{R}^{2n} \times \mathbb{R}^{2r}$ such that

$$
\psi = \frac{\epsilon^{-1}z + i\zeta}{\gamma\sqrt{2}}, \qquad \eta = \dot{y}.
\tag{III.8}
$$

Now, the coupling equations (III.4) transform into the system

$$
\begin{aligned}
\dot{y}_\epsilon &= \eta_\epsilon, & \dot{\eta}^j_\epsilon &= -\tfrac{1}{2}\langle \partial_j H(y_\epsilon)\zeta_\epsilon, \zeta_\epsilon \rangle - \tfrac{1}{2}\epsilon^{-2}\langle \partial_j H(y_\epsilon)z_\epsilon, z_\epsilon \rangle \\
\dot{z}_\epsilon &= H(y_\epsilon)\zeta_\epsilon, & \dot{\zeta}_\epsilon &= -\epsilon^{-2}H(y_\epsilon)z_\epsilon.
\end{aligned}
$$

---

[52]The reader should be aware that there are slight changes of notation compared to Chapter II.

[53]This assumption simplifies the following discussion. It is by no means essential, however, as the general results of Chapter IV will show.

A simple calculation reveals that this system just is the canonical system[54]

$$\dot{y}_\epsilon = \frac{\partial E_\epsilon}{\partial \eta}, \quad \dot{\eta}_\epsilon = -\frac{\partial E_\epsilon}{\partial y}, \quad \dot{z}_\epsilon = \frac{\partial E_\epsilon}{\partial \zeta}, \quad \dot{\zeta}_\epsilon = -\frac{\partial E_\epsilon}{\partial z}, \qquad \text{(III.9)}$$

belonging to the energy function

$$E_\epsilon = \tfrac{1}{2}|\eta|^2 + \tfrac{1}{2}\langle H(y)\zeta, \zeta\rangle + \tfrac{1}{2}\epsilon^{-2}\langle H(y)z, z\rangle = \tfrac{1}{2}|\dot{y}|^2 + \langle H(y)\psi, \psi\rangle.$$

Thus, η is the conjugate momentum for y, and ζ for z. We now switch to Lagrangian mechanics by applying a Legendre transform, [1, Chap. 3.6], to the energy function E_ϵ. The configuration space of the obtained Lagrangian $\mathscr{L}_\epsilon$ will be $M = \mathbb{R}^n \times \mathbb{R}^r$. Now, the Legendre transform yields

$$\mathscr{L}_\epsilon(y, z, \dot{y}, \dot{z}) = \langle \dot{y}, \eta\rangle + \langle \dot{z}, \zeta\rangle - E_\epsilon, \qquad \dot{y} = \frac{\partial E_\epsilon}{\partial \eta}, \quad \dot{z} = \frac{\partial E_\epsilon}{\partial \zeta}.$$

Evaluating these expression leads to

$$\zeta = H(y)^{-1}\dot{z},$$

and, with $x = (y, z)$, to the Lagrangian

$$\mathscr{L}_\epsilon(x, \dot{x}) = \tfrac{1}{2}\langle \dot{x}, \dot{x}\rangle_G - \epsilon^{-2}U(x), \qquad \dot{x} \in T_x M. \qquad \text{(III.10)}$$

Here, the Riemannian metric $\langle \cdot, \cdot\rangle_G$ on the configuration space M is given by the metric tensor

$$G(y, z) = \begin{pmatrix} I & 0 \\ 0 & H(y)^{-1} \end{pmatrix}$$

and the potential U by the function

$$U(y, z) = \tfrac{1}{2}\langle H(y)z, z\rangle.$$

The critical manifold $N = \{x \in M : U(x) = 0\}$ of this *nonnegative* potential is just

$$N = \mathbb{R}^n \times \{0\} \subset M.$$

On N, the Hessian H_U of the potential U, as defined[55] in §II.1.1, is given by

$$H_U(y) = G^{-1}(x)D^2U(x)\big|_{x=(y,0)} = \begin{pmatrix} 0 & 0 \\ 0 & H^2(y) \end{pmatrix}.$$

Assumption (III.5) is equivalent to the nondegeneracy condition, Eq. (II.6), showing that the potential U is *constraining* to the critical submanifold N in

---

[54]This canonical structure of the quantum-classical-coupling model has been introduced by BORNEMANN, NETTESHEIM, and SCHÜTTE [17, §IV.C].

[55]The reader should notice that in Chapter II the Hessian H_U was simply denoted by the letter H. This should not be confused with the *Hamiltonian* H here.

the sense of Definition II.1. Now, we proceed further along the lines of §II.1. We assume that the Hamiltonian H has a smooth spectral decomposition

$$H(y) = \sum_{\lambda=1}^{s} \omega_\lambda(y) P_\lambda^H(y), \qquad s \le r.$$

This yields a smooth spectral decomposition of the Hessian H_U,

$$H_U(y) = \sum_{\lambda=1}^{s} \omega_\lambda^2(y) P_\lambda(y), \qquad P_\lambda = \begin{pmatrix} 0 & 0 \\ 0 & P_\lambda^H \end{pmatrix}.$$

For calculating the homogenized potential U_{hom} of Definition II.4, we need to identify the initial values of our system in the $x = (y, z)$ coordinates. By assumption (III.7), we obtain

$$x_* = x_\epsilon(0) = (y_*, 0) \in N, \qquad v_* = \dot{x}_\epsilon(0) = (w_*, -i\gamma\sqrt{2}H(y_*)\psi_*).$$

Thus, we get the action constants

$$\theta_*^\lambda = \frac{\langle P_\lambda(y_*)v_*, P_\lambda(y_*)v_*\rangle_G}{2\,\omega_\lambda(y_*)} = \langle P_\lambda^H(y_*)\psi_*, \psi_*\rangle,$$

and the homogenized potential

$$U_{\text{hom}}(y) = \sum_\lambda \theta_*^\lambda \,\omega_\lambda(y).$$

The equation of motion belonging to the corresponding natural mechanical system on N are given by

$$\ddot{y}_{\text{hom}}^j = -\partial_j U_{\text{hom}}(y_{\text{hom}}), \qquad j = 1, \ldots, n, \tag{III.11}$$

with initial values $y_{\text{hom}}(0) = y_*$ and $\dot{y}_{\text{hom}}(0) = w_*$. This is the finite-dimensional analog of what is known as the *time-dependent Born-Oppenheimer* model in quantum mechanics [22][41], also known as the *quantum adiabatic approximation*. We will come back to that point in Chapter IV.

Theorem II.1 and its proof in §II.2 yield the following result.[56]

Theorem 1. *Let the Born-Oppenheimer solution y_{hom} be non-flatly resonant up to order three. Then, for every finite time interval $[0, T]$ we obtain the strong convergence*

$$y_\epsilon \to y_{\text{hom}} \quad \text{in} \quad C^1([0, T], \mathbb{R}^n)$$

and the weak* convergence

$$\psi_\epsilon \overset{*}{\rightharpoonup} 0 \quad \text{in} \quad L^\infty([0, T], \mathbb{C}^r).$$

---

[56] An asymptotic study yielding error estimates is subject of Appendix C. There, however, more restrictive resonance assumptions have to be applied.

Proof. Conservation of energy shows that $\dot{y}_\epsilon$ is uniformly bounded. Thus, on the finite time interval $[0, T]$, we obtain uniform bounds for y, z, and ζ. We may thus restrict ourselves to compact portions of the energy surface and the critical submanifold N. Then, Theorem II.1 is applicable and yields the asserted strong convergence $y_\epsilon \to y_{\text{hom}}$ in C^1. Putting all coordinates in the right order, we obtain

$$\psi_\epsilon = \frac{\epsilon^{-1} z_\epsilon + i H(y_\epsilon)^{-1} \dot{z}_\epsilon}{\gamma \sqrt{2}}.$$

Now, the weak* convergence $\psi_\epsilon \overset{*}{\rightharpoonup} 0$ follows from Lemmas II.2 and II.4. $\square$

Remark 3. This Theorem still holds true, if we relax the resonance condition by only assuming that y_{hom} is non-flatly resonant up to order *two*. The reasons for this are the special structure of the metric tensor G and the potential U being *quadratic*. Now, a careful review of the proof of Theorem II.1 would reveal that those terms that we have shown to converge weakly* to zero because of the third-order resonance conditions, are *identically zero* in this particular case. However, we have refrained from formulating this fact explicitly, since we will prove a far more general result in Chapter IV.

A simple calculation shows the conservation of the Euclidean norm of the "quantum" state,

$$|\psi_\epsilon(t)| = |\psi_*|.$$

Also, this follows from more general conservation properties which we will discuss in Chapter IV. Thus, the weak* convergence $\psi_\epsilon \overset{*}{\rightharpoonup} 0$ is strong *if and only if* we consider the trivial initial value $\psi_* = 0$, which is physically not interesting. This is in perfect accordance with Lemma II.17.

Adiabatic Results in Quantum Theory and Quantum-Classical Coupling

Most commonly, the simulation of the dynamical behavior of molecular systems is based on the assumption that the system of interest can sufficiently well be described by models of classical mechanics, cf. §III.2. However, such classical molecular dynamics approaches cannot be valid if the very nature of the process under consideration is *quantum mechanically*: e.g., the transfer of key protons in enzymes, clusters, or matrices. In all these cases a quantum dynamical description is unavoidable. Since a full quantum dynamics simulation of, e.g., a complete enzyme is not feasible, so-called *mixed quantum-classical* models have found growing interest in applications. These models describe most atoms by the means of classical mechanics but an important, small portion of the underlying system by the means of quantum mechanics.

In the current literature various mixed quantum-classical models have been proposed. We will restrict our attention to a particular model, the so-called QCMD (quantum-classical molecular dynamics) model, which has been used extensively for real life applications, cf. [9][17] and the references cited therein. Our concern is a further mathematical understanding of this model by studying its singular limit. Besides yielding analytical insight into the model, this study opens the way towards advanced numerical techniques.

For the sake of simplicity, we introduce this coupling model in the case of two particles. We assume that they have spatial coordinates $x \in \mathbb{R}^d$ and $y \in \mathbb{R}^n$, with mass $m = \epsilon^2 \ll 1$, respectively $M = 1$. The interaction potential will be denoted by $V(y,x)$. The lighter particle is supposed to perform a quantum motion. It thus has to be described by a quantum Hamiltonian H which is typically of the form

$$H(y) = -\Delta_x + V(y,x),$$

where Δ_x denotes the Laplacian with respect to x. Hence, the Hamiltonian is parametrized by the position y of the heavier particle, the description of which remains classical. The equations of motion of the coupling model are given by

$$\ddot{y}_\epsilon^j = -\langle \partial_j H(y_\epsilon)\psi_\epsilon, \psi_\epsilon \rangle, \qquad j = 1, \ldots, n,$$

$$i\epsilon\dot{\psi}_\epsilon = H(y_\epsilon)\psi_\epsilon.$$

Here, $\langle \cdot, \cdot \rangle$ denotes the inner product of the Hilbert space $\mathscr{H} = L^2(\mathbb{R}^d)$ under consideration.

The singular limit of the coupling model can be put into the broader framework of *adiabatic results* in quantum theory. In fact, the limit equation that we will obtain can be motivated by referring to the *adiabatic theorem* of quantum mechanics. The adiabatic theorem will be subject of §1 and, based upon it, the mentioned motivation of the limit equation will be given at the beginning of §2, where we address the singular limit of the coupling model.

Recall that we have studied an analogue of the coupling model with a Hamiltonian operating on a *finite* dimensional Hilbert space $\mathscr{H}$ in §III.3. There, we have shown that the coupling model can be transformed to a natural mechanical system with a strong constraining potential to which the homogenization theory of Chapter II is applicable. This result suggests that the singular limit of the coupling model, and the adiabatic theorem of quantum mechanics itself, can in principle be attacked in a more direct way by the method of weak convergence. The point is to identify the right quadratic quantity to look at. This quantity turns out to be the *density operator* $\rho_\epsilon(t)$ belonging to the wave function $\psi_\epsilon(t)$ under consideration,

$$\rho_\epsilon(t) = \langle \cdot, \psi_\epsilon(t) \rangle \cdot \psi_\epsilon(t).$$

Now, for *finite* dimensional state spaces $\mathscr{H}$, we can proceed closely along the lines of the illustrative example in §I.2. For instance, the important *weak virial theorem*, which has related the limit quadratic quantities with the Hessian in §I.2, or §II.2.2, here establishes a commutation relation between the limit density operator and the Hamiltonian, namely

$$[\rho_0, H(y_0)] = 0.$$

This way, we get a simple and, in physical terms, well-motivated proof of the finite dimensional limit result, Theorem III.1. This direct proof for finite dimensional state spaces was developed by the present author and published in joint work of BORNEMANN and SCHÜTTE [19].

The extension of this direct proof to *infinite* dimensional state spaces poses considerable functional analytical problems. *First*, and foremost, since the operator-valued functions ρ_ϵ are bounded in every reasonable respect, one would like to establish an extraction principle ensuring $\rho_\epsilon \xrightarrow{*} \rho_0$. As it turns out, such an extraction principle holds in the space

$$L^\infty([0, T], \mathscr{I}_1(\mathscr{H}))$$

of trace-class-operator-valued functions, but would not do so in the corresponding space $L^\infty([0, T], \mathscr{B}(\mathscr{H}))$ of functions having bounded operators as values.

A *second* source of problems is the fact that the Hamiltonian H is *unbounded* as an operator in $\mathscr{H}$, making it difficult to assign any direct meaning to a strong convergence $H(y_\epsilon) \to H(y_0)$ which we would like to get as a consequence of $y_\epsilon \to y_0$. However, since Hamiltonians in quantum theory are *semi-bounded* and therefore associated with a quadratic form (the energy), there is a *bounded* extension $H : \mathscr{H}_+ \to \mathscr{H}_-$, where $\mathscr{H}_+ \hookrightarrow \mathscr{H} \hookrightarrow \mathscr{H}_-$ is a so-called *rigging* of the Hilbert space $\mathscr{H}$. This way we obtain the above mentioned *strong* convergence $H(y_\epsilon) \to H(y_0)$ simply in the space $L^\infty([0,T], \mathscr{B}(\mathscr{H}_+, \mathscr{H}_-))$.

All these more sophisticated tools from functional analysis will be presented in Appendix B. Using these tools makes the proof of the adiabatic theorem in §1, and that of the singular limit result in §2, only slightly more elaborate for the infinite than for the finite dimensional case. However, the reader should shortly browse through Appendix B first, familiarizing himself with the notation that we will employ.

§1.   The Adiabatic Theorem of Quantum Mechanics

The adiabatic theorem in quantum theory refers to a situation in which the original Hamiltonian of a system is gradually changed into a new Hamiltonian. Roughly speaking, the theorem states that an eigenstate for the original energy becomes approximately an eigenstate for the new energy if the switch-on of the energy difference is sufficiently slow.

The model for this situation is given by a time-dependent Schrödinger equation with slowness parameter $\epsilon \ll 1$,

$$i\dot{\psi}_\epsilon = H(\epsilon t)\psi_\epsilon, \qquad \psi_\epsilon(0) = \psi_*.$$

The switch-on of the change takes place at time $t_0 = 0$, the switch-off at time $t_1 = T/\epsilon$. We are interested in the limit situation $\epsilon \to 0$ of an "infinitely slow" change. It is convenient to transform the time variable linearly onto the fixed interval $[0,T]$, yielding the singularly perturbed equation

$$i\epsilon\dot{\psi}_\epsilon = H(t)\psi_\epsilon, \qquad \psi_\epsilon(0) = \psi_*.$$

If $\omega(t)$ is a sufficiently smooth path in the time-dependent discrete spectrum of $H(t)$ and $P(t)$ the corresponding spectral projection, the folk theorem states that

$$\langle P(t)\psi_\epsilon(t), \psi_\epsilon(t)\rangle \to \langle P(0)\psi_*, \psi_*\rangle.$$

We will give an entirely new proof for this claim under reasonably general assumptions. A short review of the literature can be found in §1.6.

§1.1.   The Result

Let $H(t)$, $t \in [0,T]$, be a family of semi-bounded selfadjoint operators on a separable Hilbert space $\mathscr{H}$. The corresponding coercive quadratic forms[57] will be denoted by $h(t; \cdot, \cdot)$. We will make the following hypotheses:

---

[57]For this notion, and the notation we are going to use below, consult §B.2.

(H1) The form domain $\mathscr{H}_+$ of $H(t)$ is time-independent and the form h uniformly $\mathscr{H}_+$-coercive, i.e., there are constants $\gamma > 0$, and $\kappa \in \mathbb{R}$ such that

$$h(t; \psi, \psi) \geq \gamma \langle \psi, \psi \rangle_{\mathscr{H}_+} - \kappa \langle \psi, \psi \rangle$$

for all $t \in [0, T]$ and $\psi \in \mathscr{H}_+$. The corresponding rigging of $\mathscr{H}$ will be denoted by $\mathscr{H}_+ \hookrightarrow \mathscr{H} \hookrightarrow \mathscr{H}_-$.

(H2) The *bounded* extension $H(t) : \mathscr{H}_+ \to \mathscr{H}_-$ depends two times continuously differentiable on t,

$$H \in C^2([0, T], \mathscr{B}(\mathscr{H}_+, \mathscr{H}_-)).$$

(H3) There is a section $\omega \in C[0, T]$ of the discrete spectrum[58] of H,

$$\omega(t) \in \sigma_{\text{disc}}(H(t)),$$

together with a time-dependent family $P \in C^1([0, T], \mathscr{K}(\mathscr{H}))$ of orthogonal projections, such that

$$H(t)P(t) = \omega(t)P(t), \qquad t \in [0, T].$$

We assume that P is the spectral projection belonging to ω *almost everywhere*.

Notice, first, that the values of P have *finite rank*, and second, that hypothesis (H3) allows for *eigenvalue crossings* on a set of measure zero.

Theorem 1. *Suppose that the hypotheses (H1), (H2), and (H3) are valid. Then, for a given initial value $\psi_* \in \mathscr{H}_+$ with $\|\psi_*\| = 1$ and a sequence $\epsilon \to 0$ there are unique solutions*

$$\psi_\epsilon \in C([0, T], \mathscr{H}_+) \cap C^1([0, T], \mathscr{H}_-)$$

of the initial value problem

$$i\epsilon \dot{\psi}_\epsilon = H(t)\psi_\epsilon, \qquad \psi_\epsilon(0) = \psi_*.$$

The differential equation holds in the space $\mathscr{H}_-$. The solution operator is unitary, $\|\psi_\epsilon(t)\| = 1$ for all $t \in [0, T]$. As $\epsilon \to 0$, the energy level probability belonging to ω converges uniformly in time to the constant given by its initial value,

$$\langle P\psi_\epsilon, \psi_\epsilon \rangle \to \langle P(0)\psi_*, \psi_* \rangle \quad \text{in} \quad C[0, T].$$

---

[58] In quantum mechanics, one refers to points of the discrete spectrum of the Hamiltonian as *energy levels*. The well-known relation $E = \hbar\omega$ of energy and frequency allows to identify these energy levels with eigen-frequencies. For this reason, and in accordance with the notation of Chapter II and §III.3, we prefer to denote the energy levels by the letter "ω." One could also think of choosing appropriate units leading to $\hbar = 1$.

§1.2.   The Proof

The proof proceeds along the lines of the first three steps of the scheme set up in the illustrative example of §I.2.

Step 1: Energy-Principle. We start with KISYŃSKI'S nontrivial existence and uniqueness result for the time-dependent Schrödinger equation, [54, Théorème 8.1]. The proof given by this author sharpens the variational method of LIONS [64] for abstract evolution equations.

Theorem 2. (Kisyński 1963). *Suppose that hypotheses (H1) and (H2) are valid. Then, for $\epsilon > 0$ and $\psi_* \in \mathcal{H}_+$, there exists a unique solution*

$$\psi_\epsilon \in C([0,T], \mathcal{H}_+) \cap C^1([0,T], \mathcal{H}_-)$$

of the initial value problem

$$i\epsilon\dot{\psi}_\epsilon = H(t)\psi_\epsilon, \qquad \psi_\epsilon(0) = \psi_*.$$

The differential equation is understood to hold in the space $\mathcal{H}_-$. The solution operator is unitary,

$$\|\psi_\epsilon(t)\| = \|\psi_*\|, \qquad t \in [0,T].$$

If $\psi_* \in D(H(0))$, there is the additional regularity

$$\psi_\epsilon \in C^1([0,T], \mathcal{H}), \qquad \psi_\epsilon(t) \in D(H(t)), \quad t \in [0,T].$$

From now on, hypotheses (H1), (H2), and (H3) will be valid throughout. Moreover, we put $\|\psi_*\| = 1$ as usual in quantum theory. Physically reasonable, there is a uniform bound on the energy:

Lemma 1. *Let $\psi_* \in \mathcal{H}_+$. Then, as $\epsilon \to 0$, there is the uniform bound*

$$\psi_\epsilon = O(1) \quad in \quad C([0,T], \mathcal{H}_+).$$

Proof. First, assume that $\psi_* \in D(H(0))$. By the regularity result of Theorem 2, we obtain[59]

$$\frac{d}{dt}\langle H(t)\psi_\epsilon(t), \psi_\epsilon(t)\rangle$$
$$= \langle H(t)\psi_\epsilon(t), \dot{\psi}_\epsilon(t)\rangle + \langle \dot{\psi}_\epsilon(t), H(t)\psi_\epsilon(t)\rangle + \langle \dot{H}(t)\psi_\epsilon(t), \psi_\epsilon(t)\rangle$$
$$= i\epsilon^{-1}\langle H(t)\psi_\epsilon(t), H(t)\psi_\epsilon(t)\rangle - i\epsilon^{-1}\langle H(t)\psi_\epsilon(t), H(t)\psi_\epsilon(t)\rangle$$
$$\qquad\qquad + \langle \dot{H}(t)\psi_\epsilon(t), \psi_\epsilon(t)\rangle$$
$$= \langle \dot{H}(t)\psi_\epsilon(t), \psi_\epsilon(t)\rangle.$$

$$(IV.1)$$

---

[59]In physics, this result is sometimes called the "Hellmann-Feynman theorem."

Now, by hypotheses (H1) and (H2), the expression

$$\|\psi\|_t^2 = h(t; \psi, \psi) + \kappa\|\psi\|^2 = \langle H(t)\psi, \psi \rangle + \kappa\|\psi\|^2$$

defines a norm uniformly equivalent to $\|\cdot\|_{\mathscr{H}_+}$. By what be have shown for the first term of this norm expression, we obtain

$$\frac{d}{dt}\|\psi_\epsilon(t)\|_t^2 = \langle \dot{H}(t)\psi_\epsilon(t), \psi_\epsilon(t)\rangle,$$

since $\|\psi_\epsilon\|^2 \equiv 1$. Invoking hypothesis (H2), we obtain

$$\frac{d}{dt}\|\psi_\epsilon(t)\|_t^2 \leq 2c\,\|\psi_\epsilon(t)\|_t^2$$

with a constant c which is *independent* of ϵ. Thus, by the Gronwall lemma, there is the uniform bound

$$\|\psi_\epsilon(t)\|_t \leq \|\psi_*\|_0\, e^{cT}, \qquad t \in [0, T].$$

A density argument shows that this estimate still holds true if we take initial values $\psi_* \in \mathscr{H}_+$. Because of the stated uniform norm equivalence, we have proved the asserted uniform bound for the $\|\cdot\|_{\mathscr{H}_+}$-norm. □

The quadratic quantities Σ_ϵ and Π_ϵ of Chapter II generalize in the present setting to the operator-valued function

$$\rho_\epsilon = \langle \cdot, \psi_\epsilon \rangle \psi_\epsilon \ \in\ C([0,T], \mathscr{J}_1(\mathscr{H})).$$

The values of ρ_ϵ are nonnegative selfadjoint trace class operators having trace one,

$$\operatorname{tr} \rho_\epsilon(t) = \langle \psi_\epsilon(t), \psi_\epsilon(t)\rangle = 1.$$

Operators with these properties are called *density operators* in quantum theory.[60] Therefore, we call ρ_ϵ the *time-dependent density operator* associated with the pure state ψ_ϵ. Because of $\psi_\epsilon \in C([0,T], \mathscr{H}_+)$ we can extend the operator values of ρ_ϵ *uniquely* to operate on $\mathscr{H}_-$ such that

$$\rho_\epsilon \in C([0,T], \mathscr{B}(\mathscr{H}_-, \mathscr{H}_+)).$$

Now, with the help of the tools developed in §B.1, the result of uniformly bounded energy, Lemma 1, implies the following important *extraction principle*. Recall that $\sigma : \mathscr{H}_- \to \mathscr{H}_+$ denotes the Riesz representation map of the rigging of $\mathscr{H}$.

---

[60]Frequently, density operators are also called *statistical operators*. A very readable survey of the Schrödinger picture of quantum mechanics in terms of these operators was given by FANO [30], cf. also the textbook of MESSIAH [69, §VIII.21–25].

Lemma 2. *There is a subsequence of ϵ, denoted by ϵ again, and a function $\rho_0 \in L^\infty([0,T], \mathcal{B}(\mathcal{H}_-, \mathcal{H}_+))$ such that:*

$$
\begin{array}{lllll}
\text{(i)} & \rho_\epsilon \overset{*}{\rightharpoonup} \rho_0 & \text{in} & L^\infty([0,T], \mathcal{J}_1(\mathcal{H})), \\
\text{(ii)} & \rho_\epsilon \sigma^* \overset{*}{\rightharpoonup} \rho_0 \sigma^* & \text{in} & L^\infty([0,T], \mathcal{J}_1(\mathcal{H}_+)), \\
\text{(iii)} & \sigma^* \rho_\epsilon \overset{*}{\rightharpoonup} \sigma^* \rho_0 & \text{in} & L^\infty([0,T], \mathcal{J}_1(\mathcal{H}_-)), \\
\text{(iv)} & \rho_\epsilon H \overset{*}{\rightharpoonup} \rho_0 H & \text{in} & L^\infty([0,T], \mathcal{J}_1(\mathcal{H}_+)), \\
\text{(v)} & H \rho_\epsilon \overset{*}{\rightharpoonup} H \rho_0 & \text{in} & L^\infty([0,T], \mathcal{J}_1(\mathcal{H}_-)).
\end{array}
$$

Proof. The uniform boundedness of $\|\psi_\epsilon\|$ and $\|\psi_\epsilon\|_{\mathcal{H}_+}$ shows that $\{\rho_\epsilon\}$ is bounded in $L^\infty([0,T], \mathcal{J}_1(\mathcal{H}))$, $\{\rho_\epsilon \sigma^*\}$ is bounded in $L^\infty([0,T], \mathcal{J}_1(\mathcal{H}_+))$, and $\{\sigma^* \rho_\epsilon\}$ is bounded in $L^\infty([0,T], \mathcal{J}_1(\mathcal{H}_-))$. By Theorem B.3 and Lemma B.10, this immediately implies the existence of a limit function $\rho_0 \in L^\infty([0,T], \mathcal{B}(\mathcal{H}_-, \mathcal{H}_+))$ and a subsequence such that the assertions (i), (ii), and (iii) hold. By hypothesis (H2) we have

$$H\sigma \in C([0,T], \mathcal{B}(\mathcal{H}_-)), \qquad \sigma H \in C([0,T], \mathcal{B}(\mathcal{H}_+)).$$

Hence, Lemma B.4 shows that

$$\rho_\epsilon H = \rho_\epsilon \sigma^* \cdot \sigma H \overset{*}{\rightharpoonup} \rho_0 \sigma^* \cdot \sigma H = \rho_0 H \quad \text{in} \quad L^\infty([0,T], \mathcal{J}_1(\mathcal{H}_+)),$$

proving assertion (iv). Likewise, we obtain

$$H\rho_\epsilon = H\sigma \cdot \sigma^* \rho_\epsilon \overset{*}{\rightharpoonup} H\sigma \cdot \sigma^* \rho_0 = H\rho_0 \quad \text{in} \quad L^\infty([0,T], \mathcal{J}_1(\mathcal{H}_-)),$$

proving assertion (v). ◻

Step 2: The Weak Virial Theorem. As in the illustrative example of §I.2, the weak limit relations of Lemma 2 allow to prove a significant commutation relation.

Lemma 3. (Weak Virial Theorem). *There holds the commutation relation*

$$\rho_0(t)\, H(t) \subset H(t)\, \rho_0(t)$$

for almost all $t \in [0,T]$ as unbounded, densely defined operators in $\mathcal{H}$.

Proof. Theorem 2 shows that

$$\rho_\epsilon^\sigma = \langle \cdot, \psi_\epsilon \rangle_{\mathcal{H}_-} \psi_\epsilon \in C^1([0,T], \mathcal{J}_1(\mathcal{H}_-)).$$

Since we have conservation of norm, $\|\psi_\epsilon\| = 1$, the sequence ρ_ϵ^σ is uniformly bounded in $C([0,T], \mathcal{J}_1(\mathcal{H}_-))$, yielding

$$i\epsilon \rho_\epsilon^\sigma \to 0 \quad \text{in} \quad C([0,T], \mathcal{J}_1(\mathcal{H}_-)).$$

Using the notation introduced in Lemma B.8, we get the time derivative

$$
\begin{aligned}
i\epsilon\dot{\rho}_\epsilon^\sigma &= -\langle\,\cdot\,,i\epsilon\dot{\psi}_\epsilon\rangle_{\mathscr{H}_-}\psi_\epsilon + \langle\,\cdot\,,\psi_\epsilon\rangle_{\mathscr{H}_-}\,i\epsilon\dot{\psi}_\epsilon \\
&= -\langle\,\cdot\,,H\psi_\epsilon\rangle_{\mathscr{H}_-}\psi_\epsilon + \langle\,\cdot\,,\psi_\epsilon\rangle_{\mathscr{H}_-}H\psi_\epsilon \\
&= (H\rho_\epsilon)^\sigma - (\rho_\epsilon H)^\sigma.
\end{aligned}
$$

Taking weak* limits in $L^\infty([0,T],\,\mathscr{J}_1(\mathscr{H}_-))$ shows, by applying Lemma 2, Lemma B.9, and Lemma B.3, that

$$
0 = (H\rho_0)^\sigma - (\rho_0 H)^\sigma.
$$

Now, given an arbitrary element $\psi \in \mathscr{H}_+$, we get

$$
0 = (H(t)\rho_0(t))^\sigma\,\sigma^*\psi - (\rho_0(t)H(t))^\sigma\,\sigma^*\psi = H(t)\rho_0(t)\psi - \rho_0(t)H(t)\psi,\quad (*)
$$

for almost all $t \in [0,T]$. Since $\rho_0(t)\psi \in \mathscr{H}_+$ and $\rho_0(t)H(t)\psi \in \mathscr{H}_+$, we thus have shown that

$$
\rho_0(t)D(H(t)) \subset \rho_0(t)\mathscr{H}_+ \subset D(H(t)),\qquad (**)
$$

almost everywhere. Combining (*) and (**) gives the assertion.   □

Remark 1. On a first sight, calling Lemma 3 the "weak virial theorem" might appear a little bit strange. However, this is justified by the analogy in proof to Lemma II.7. Moreover, there is a strong analogy between the virial theorem in classical mechanics and results in quantum theory which state that the *expectation value* of certain commutators is *zero*. This analogy was first studied by HIRSCHFELDER [46], who coined the notion *hypervirial theorem* for such a kind of result. These hypervirial theorems play a prominent role in modern quantum chemistry, cf. the monograph of FERNÁNDEZ and CASTRO [31]. As in the case of classical mechanics, §I.2.6, we replace the average, i.e., the expectation value, by the weak*-limit: the proof of Lemma 3 has shown that $[H,\rho_\epsilon] \xrightarrow{*} 0$, if interpreted in the correct way.

As an immediate corollary we obtain the following lemma which has been the ultimate goal of this step of the proof.

Lemma 4. *For almost all $t \in [0,T]$ there holds the commutation relation*

$$
[\rho_0(t),P(t)] = 0 \quad in \quad \mathscr{B}(\mathscr{H}).
$$

Proof. The spectral theorem for unbounded selfadjoint operators, e.g., in the very formulation as Theorem 13.33 in RUDIN'S textbook [83], teaches the following: If $P(t)$ is a spectral projection of $H(t)$ and the *bounded* operator $\rho_0(t)$ commutes with $H(t)$ in the sense that $\rho_0(t)H(t) \subset H(t)\rho_0(t)$, then one obtains

$$P(t)\rho_0(t) = \rho_0(t)P(t). \qquad (*)$$

By hypothesis (H3), $P(t)$ is the spectral projection belonging to $\omega(t)$ almost everywhere. By Lemma 3, $\rho_0(t)$ commutes with $H(t)$ in the appropriate sense almost everywhere. Thus, (*) holds almost everywhere. □

Step 3: Adiabatic Invariance of the Action. According to the correspondence principle of EHRENFEST, the "action" of the quantum mechanical system under consideration is given by the (time-dependent) energy level probability

$$\theta_\epsilon = \langle P\psi_\epsilon, \psi_\epsilon \rangle \in C^1[0,T].$$

At time $t = 0$, the energy level probability is given by the ϵ-independent value

$$\theta_\epsilon(0) = \theta_*, \qquad \theta_* = \langle P(0)\psi_*, \psi_* \rangle.$$

The proof of Theorem 1 is finished by the following lemma.

Lemma 5. *For the originally given sequence $\epsilon \to 0$, there is*

$$\theta_\epsilon \to \theta_* = \text{const} \quad in \quad C[0,T].$$

Proof. First, let ϵ be the *subsequence* which has been introduced in Lemma 2. Since $P \in C^1([0,T], \mathcal{K}(\mathcal{H}))$, we obtain by Lemma B.5 that

$$\theta_\epsilon = \text{tr}(P\rho_\epsilon) \xrightarrow{*} \text{tr}(P\rho_0) = \theta_0 \quad in \quad L^\infty[0,T]. \qquad (IV.2)$$

Recall that $P(t)$ is the orthogonal projection onto a *finite dimensional* subspace of $D(H(t)) \subset \mathcal{H}_+$. Hence, the time derivative of θ_ϵ is given by

$$
\begin{aligned}
\dot\theta_\epsilon &= \langle P\psi_\epsilon, \dot\psi_\epsilon \rangle + \langle \dot\psi_\epsilon, P\psi_\epsilon \rangle + \langle \dot P\psi_\epsilon, \psi_\epsilon \rangle \\
&= i\epsilon^{-1}\langle P\psi_\epsilon, H\psi_\epsilon \rangle - i\epsilon^{-1}\langle H\psi_\epsilon, P\psi_\epsilon \rangle + \langle \dot P\psi_\epsilon, \psi_\epsilon \rangle \\
&= i\epsilon^{-1}\langle HP\psi_\epsilon, \psi_\epsilon \rangle - i\epsilon^{-1}\langle \psi_\epsilon, HP\psi_\epsilon \rangle + \langle \dot P\psi_\epsilon, \psi_\epsilon \rangle \\
&= i\epsilon^{-1}\omega\langle P\psi_\epsilon, \psi_\epsilon \rangle - i\epsilon^{-1}\omega\langle \psi_\epsilon, P\psi_\epsilon \rangle + \langle \dot P\psi_\epsilon, \psi_\epsilon \rangle \\
&= \langle \dot P\psi_\epsilon, \psi_\epsilon \rangle = \text{tr}(\dot P\rho_\epsilon).
\end{aligned}
$$

Employing Lemma B.5 once again, we conclude that

$$\dot\theta_\epsilon = \text{tr}(\dot P\rho_\epsilon) \xrightarrow{*} \text{tr}(\dot P\rho_0) = \dot\theta_0 \quad in \quad L^\infty[0,T].$$

In particular, the sequence $\{\dot\theta_\epsilon\}$ is uniformly bounded in $L^\infty[0,T]$. The extended Arzelà-Ascoli theorem, Principle I.4, implies therefore the uniform convergence

$$\theta_\epsilon \to \theta_0 \quad in \quad C[0,T]$$

and the regularity $\theta_0 \in C^{0,1}[0,T]$. Now, using the commutativity result, Lemma 4, we obtain

$$\begin{aligned}
\dot{\theta}_0 &= \operatorname{tr}(\dot{P}\rho_0) = \operatorname{tr}(P\dot{P} \cdot \rho_0 \cdot P) + \operatorname{tr}((I-P)\dot{P} \cdot \rho_0 \cdot (I-P)) \\
&= \operatorname{tr}(P\dot{P}P \cdot \rho_0) + \operatorname{tr}((I-P)\dot{P}(I-P) \cdot \rho_0) = 0,
\end{aligned}$$

almost everywhere. Recall that there holds $P\dot{P}P = (I-P)\dot{P}(I-P) = 0$ by Lemma II.9. Thus, we get $\theta_0 \equiv \theta_0(0) = \theta_*$. Since this limit is independent of the subsequence chosen, we can discard the extraction of subsequences by recalling Principle I.5.   □

§1.3.   Extension to the Essential Spectrum

So far we have proven the adiabatic invariance only for energy level probabilities which belong to the *discrete* spectrum of H. However, since the solution operator is *unitary*, we can prove similar results for the essential spectrum as a *whole*.

To this end, we replace hypothesis (H3) by:

(H3′) Denote by $Q(t)$ the spectral projection of $H(t)$ belonging to the essential spectrum $\sigma_{\mathrm{ess}}(H(t))$. There is a number $N_{\mathrm{disc}} \in \mathbb{N}_0 \cup \{\infty\}$ and a family $\omega_\lambda \in C[0,T]$ which builds up the discrete spectrum of H,

$$\{\omega_\lambda(t)\}_{\lambda=1}^{N_{\mathrm{disc}}} = \sigma_{\mathrm{disc}}(H(t)).$$

Further, there is a family $P_\lambda \in C^1([0,T], \mathscr{K}(\mathscr{H}))$ the values of which are orthogonal projections obeying

$$P_\lambda(t)Q(t) = Q(t)P_\lambda(t) = 0, \qquad \operatorname{range} P_\lambda(t) \subset D(H(t)),$$

and

$$H(t)P_\lambda(t) = \omega_\lambda(t)P_\lambda(t), \qquad \lambda = 1,\ldots,N_{\mathrm{disc}}.$$

In the strong operator topology, all these projections sum up to the identity operator,

$$I = Q(t) + \sum_\lambda P_\lambda(t), \qquad t \in [0,T].$$

The set of resonance points,

$$I_r = \{t \in [0,T] : \omega_\lambda(t) = \omega_\mu(t) \text{ for some } \lambda \neq \mu\},$$

is of measure zero.

Notice that hypothesis (H3′) implies the validity of hypothesis (H3) for every pair $(\omega_\lambda, P_\lambda)$. We define the energy level probabilities

$$\theta_\epsilon^\lambda = \langle P_\lambda \psi_\epsilon, \psi_\epsilon \rangle, \qquad \theta_\epsilon^Q = \langle Q\psi_\epsilon, \psi_\epsilon \rangle.$$

Under hypothesis (H1), (H2), and (H3′), Theorem 1 shows that

$$\theta_\epsilon^\lambda \to \theta_*^\lambda = \langle P_\lambda(0)\psi_*, \psi_* \rangle \quad \text{in} \quad C[0,T]. \tag{IV.3}$$

However, we *cannot* prove the same way that

$$\theta_\epsilon^Q \to \theta_*^Q = \langle Q(0)\psi_*, \psi_* \rangle \quad \text{in} \quad C[0,T], \tag{IV.4}$$

for the following reason: despite the fact that the assertion of Lemma 4 applies to $Q(t)$ with literally the same proof, i.e., there is

$$[\rho_0(t), Q(t)] = 0, \tag{IV.5}$$

for almost all $t \in [0,T]$, the proof of Lemma 5 does not work any longer because of

$$Q(t) \notin \mathcal{K}(\mathcal{H}).$$

For $Q(t)$ is a orthogonal projection of *infinite* rank.

However, there are two situations where we can prove a convergence like Eq. (IV.4). This possibility is based on the observation that the solution operator being unitary implies

$$\theta_\epsilon^Q(t) + \sum_\lambda \theta_\epsilon^\lambda(t) = \|\psi_\epsilon(t)\|^2 = 1 = \|\psi_*\|^2 = \theta_*^Q + \sum_\lambda \theta_*^\lambda. \tag{IV.6}$$

First, completely trivial, there is the following corollary of Theorem 1.

Corollary 1. *Let hypotheses (H1), (H2), and (H3′) be valid. If the discrete spectrum is finite, $N_{\mathrm{disc}} < \infty$, then $\theta_\epsilon^Q \to \theta_*^Q$ in $C[0,T]$.*

Second, and far less trivial, there is the following corollary of Theorem 1.

Corollary 2. *Let hypotheses (H1), (H2), and (H3′) be valid. If the essential spectrum is not excited initially, i.e., $Q(0)\psi_* = 0$, then there is the convergence*

$$\theta_\epsilon^Q(t) \to \theta_*^Q = 0,$$

pointwise for all $t \in [0,T]$.

Proof. The isometry relation (IV.6) shows that

$$\|\{\theta_\epsilon^\lambda(t)\}_\lambda\|_{\ell^1} = \sum_\lambda \theta_\epsilon^\lambda(t) \leq 1 = \sum_\lambda \theta_*^\lambda = \|\{\theta_*^\lambda\}_\lambda\|_{\ell^1}.$$

Thus, the component-wise convergence Eq. (IV.3) and Lemma 6 below yield the pointwise convergence

$$\sum_\lambda \theta_\epsilon^\lambda(t) \to \sum_\lambda \theta_*^\lambda = 1, \quad t \in [0,T],$$

which implies the assertion. □

Lemma 6. *Let there be a sequence $\{x_\epsilon\}$ in the space ℓ^1. Suppose that this sequence converges component-wise to an element $x_0 \in \ell^1$,*

$$x_\epsilon(\lambda) \to x_0(\lambda), \qquad \lambda \in \mathbb{N},$$

while the norm remains bounded as $\|x_\epsilon\|_{\ell^1} \leq \|x_0\|_{\ell^1}$. Then, there is the strong convergence

$$x_\epsilon \to x_0 \quad \text{in} \quad \ell^1.$$

Proof. The component-wise convergence and the uniform bound in norm imply that $x_\epsilon \overset{*}{\rightharpoonup} x_0$ in $\ell^1 = c_0^*$. Hence, we have

$$\|x_0\|_{\ell^1} \leq \liminf \|x_\epsilon\|_{\ell^1} \leq \limsup \|x_\epsilon\|_{\ell^1} \leq \|x_0\|_{\ell^1},$$

i.e., $\|x_0\|_{\ell^1} = \lim \|x_\epsilon\|_{\ell^1}$. By a well-known result about the space ℓ^1, [45, Theorem 13.47], this convergence of the norm, together with the assumption of component-wise convergence, implies the strong convergence in ℓ^1.
□

§1.4. Remarks on the Limit of the Time-Dependent Density Operator

Here, we study the limit function ρ_0 of Lemma 2 more closely. The results of Lemma B.6 show that the values $\rho_0(t)$ are nonnegative selfadjoint trace class operators, almost everywhere. However, the weak* convergence

$$\rho_\epsilon \overset{*}{\rightharpoonup} \rho_0 \quad \text{in} \quad L^\infty([0,T], \mathscr{J}_1(\mathscr{H}))$$

does *not suffice* to conclude that

$$1 \equiv \operatorname{tr}\rho_\epsilon \overset{*}{\rightharpoonup} \operatorname{tr}\rho_0 \quad \text{in} \quad L^\infty[0,T],$$

and therefore, that ρ_0 has trace one. In fact, even though the trace map

$$\operatorname{tr} : L^\infty([0,T], \mathscr{J}_1(\mathscr{H})) \to L^\infty([0,T],\mathbb{C})$$

is norm continuous, it is not weakly* sequentially continuous, as shown in §B.1.4. Thus, without further structure, we *cannot* state that the values $\rho_0(t)$ are density operators almost everywhere. Instead, one can only show the following estimate:

$$
\begin{aligned}
\operatorname{tr}\rho_0(t) \;\; &\leq \;\; \|\operatorname{tr}\rho_0\|_{L^\infty[0,T]} = \|\rho_0\|_{L^\infty([0,T],\mathscr{J}_1(\mathscr{H}))} \\
&\leq \;\; \liminf \|\rho_\epsilon\|_{L^\infty([0,T],\mathscr{J}_1(\mathscr{H}))} \qquad\qquad\text{(IV.7)}\\
&= \;\; \liminf \|\operatorname{tr}\rho_\epsilon\|_{L^\infty[0,T]} = 1,
\end{aligned}
$$

almost everywhere. However, in the setting of Corollary 2 one can actually prove more.

Corollary 3. *Let hypotheses (H1), (H2), and (H3') be valid. If $Q(0)\psi_* = 0$, one obtains*

$$\operatorname{tr}\rho_0 = 1 \quad \text{in} \quad L^\infty[0,T], \qquad \rho_0 Q = Q\rho_0 Q = 0 \quad \text{in} \quad L^\infty([0,T], \mathscr{I}_1(\mathscr{H})).$$

Proof. Eq. (IV.2) shows that $\theta_*^\lambda = \theta_0^\lambda = \operatorname{tr}(P_\lambda \rho_0)$. Thus, Eqs. (IV.6) and (IV.7) yield that

$$1 = \sum_\lambda \theta_*^\lambda \leq \operatorname{tr}(Q(t)\rho_0(t)) + \sum_\lambda \operatorname{tr}(P_\lambda(t)\rho_0(t)) = \operatorname{tr}\rho_0(t) \leq 1,$$

almost everywhere, which is equivalent to $\operatorname{tr}\rho_0 = 1$ in $L^\infty[0,T]$. Moreover, this two-sided estimate shows that

$$0 = \operatorname{tr}(Q(t)\rho_0(t)) = \operatorname{tr}(Q(t)\rho_0(t)Q(t)) = \|Q(t)\rho_0(t)Q(t)\|_{\mathscr{I}_1(\mathscr{H})}, \quad \text{a.e.,}$$

since $Q(t)\rho_0(t)Q(t)$ is nonnegative. Thus, by using the commutativity relation Eq. (IV.5), we obtain $\rho_0 Q = \rho_0 Q^2 = Q\rho_0 Q = 0$. ☐

Up to now, the limit function ρ_0 cannot be identified *uniquely*, i.e., independent of the defining subsequence. However, under stronger assumptions, the limit ρ_0 can be described with the help of the initial data.

Corollary 4. *Let hypotheses (H1), (H2), and (H3') be valid. Suppose that $Q(0)\psi_* = 0$ and*

$$\operatorname{rank} P_\lambda(0) > 1 \implies P_\lambda(0)\psi_* = 0.$$

Then, for the originally given sequence $\epsilon \to 0$, one obtains

$$\rho_\epsilon \overset{*}{\rightharpoonup} \rho_0 = \sum_\lambda \theta_*^\lambda P_\lambda \quad \text{in} \quad L^\infty([0,T], \mathscr{I}_1(\mathscr{H})),$$

the sum being understood to converge in the strong operator topology.

Proof. Lemma 4 and the corresponding result for Q, Eq. (IV.5), show that

$$\rho_0 = Q\rho_0 Q + \sum_\lambda P_\lambda \rho_0 P_\lambda, \qquad (*)$$

almost everywhere. Corollary 3 yields $Q\rho_0 Q = 0$. On the other hand, there is

$$\|P_\lambda(t)\rho_0(t)P_\lambda(t)\|_{\mathscr{I}_1} = \operatorname{tr}(P_\lambda(t)\rho_0(t)) = \theta_0^\lambda = \theta_*^\lambda,$$

almost everywhere. Thus, since θ_*^λ is nonzero for projections $P_\lambda(t)$ of rank one only, we obtain

$$P_\lambda \rho_0 P_\lambda = \theta_*^\lambda P_\lambda, \qquad \lambda = 1, \ldots, N_{\text{disc}}.$$

Summarizing, the limit $(*)$ is given by the asserted expression, *independently* of the defining subsequence. Therefore, by Principle I.5, we may discard the extraction of subsequences. ☐

This way, the limit density operators $\rho_0(t)$ are *convex combinations* of density operators of rank one. In particular, the rank of $\rho_0(t)$ is given by

$$\operatorname{rank}\rho_0(t) = \#\{\lambda : \theta_*^\lambda \neq 0\},$$

almost everywhere.

§1.5. The Example of a Two-Body Hamiltonian

As an example for the abstract results of the preceding sections, we consider a time-dependent two-body Hamiltonian on $\mathcal{H} = L^2(\mathbb{R}^3)$,

$$H(t) = -\Delta + V(t, \cdot),$$

where the potential fulfills

$$V = V_1 + V_2, \qquad V_1 \in C^2([0,T], \mathcal{R}), \qquad V_2 \in C^2([0,T], L^\infty(\mathbb{R}^3)).$$

The discussion of the *Rollnik class* $\mathcal{R}$ in §B.2.1 shows that the hypotheses (H1) and (H2) of §1.1 are valid. The associated rigging of $\mathcal{H} = L^2(\mathbb{R}^3)$ is given by the scale of Sobolev spaces,

$$\mathcal{H}_- = H^{-1}(\mathbb{R}^3), \qquad \mathcal{H}_+ = H^1(\mathbb{R}^3).$$

Assuming hypothesis (H3′), Theorem 1 yields the adiabatic invariance

$$\langle P_\lambda \psi_\epsilon, \psi_\epsilon \rangle \;\rightarrow\; \langle P_\lambda(0)\psi_*, \psi_* \rangle \quad \text{in} \quad C[0,T], \quad \lambda = 1, \ldots, N_{\text{disc}},$$

of the energy level probabilities for the *discrete spectrum*. Under the additional assumption that the discrete spectrum is finite, $N_{\text{disc}} < \infty$, or the essential spectrum is not excited initially, $Q(0)\psi_* = 0$, Corollaries 1 and 2 yield the adiabatic invariance of the energy level probability belonging to the *essential spectrum* as a whole,

$$\langle Q(t)\psi_\epsilon(t), \psi_\epsilon(t) \rangle \;\rightarrow\; \langle Q(0)\psi_*, \psi_* \rangle, \qquad t \in [0,T].$$

Looking at these results, three questions arise naturally: first, whether the discrete spectrum is finite, i.e., $N_{\text{disc}} < \infty$, second, whether σ_{ess} is connected, and third, whether $\sigma_{\text{ess}} = \emptyset$. For the convenience of the reader, we recall some criteria from REED'S and SIMON'S textbook [77] which help to decide about these questions:

Denote by $\mathcal{R} + L^\infty(\mathbb{R}^3)_\delta$ the space

$$\mathcal{R} + L^\infty(\mathbb{R}^3)_\delta \;=\; \{V : \forall \delta > 0 \; \exists V = V_{1\delta} + V_{2\delta},$$

$$V_{1\delta} \in \mathcal{R}, \; V_{2\delta} \in L^\infty(\mathbb{R}^3), \; \|V_{2\delta}\|_{L^\infty} \leq \delta\}.$$

Suppose that $V(t, \cdot) \in \mathcal{R} + L^\infty(\mathbb{R}^3)_\delta$ for all $t \in [0,T]$. Then, as an application of Weyl's theorem, Example XIII.4.7 of [77] teaches that the essential spectrum is connected, in fact

$$\sigma_{\text{ess}}(H(t)) = [0, \infty[, \qquad t \in [0,T].$$

Additionally, Theorem XIII.6 of [77] states the following:

(a) Suppose that

$$V(0, x) \leq -a|x|^{-2+\eta}, \qquad |x| > R_0,$$

for some R_0 and some $a > 0$, $\eta > 0$. Then the discrete spectrum is infinite, $N_{\text{disc}} = \infty$.

(b) Suppose that

$$V(0, x) \geq -\tfrac{1}{4}b|x|^{-2}, \qquad |x| > R_0,$$

for some R_0 and some $b < 1$. Then the discrete spectrum is finite, $N_{\text{disc}} < \infty$.

If it happens to be that $V(0, \cdot) \in \mathscr{R}$, then the so-called Birman-Schwinger bound, [77, Theorem XIII.10], states that

$$N_{\text{disc}} \leq \left(\frac{\|V(0, \cdot)\|_{\mathscr{R}}}{4\pi} \right)^2 < \infty.$$

On the other hand, Theorem XIII.16 of [77] tells us that $H(t)$ has a purely discrete spectrum, i.e.,

$$\sigma_{\text{ess}}(H(t)) = \emptyset,$$

if $V(t, \cdot)$ is a locally bounded nonnegative function such that $V(t, x) \to \infty$ as $|x| \to \infty$.

§1.6. Bibliographical Remarks

Our approach to the adiabatic theorem of quantum mechanics is somewhat complementary to the existing literature. By looking at density operators instead of the wave function we directly attack the energy level probabilities. This way, however, we have stated and proven nothing about the convergence of the solution ψ_ϵ itself. In this respect, our result is *weaker* than what can be found in the literature. On the other hand, our result is *stronger*, since we do not have to employ a so-called "gap condition" for the discrete part of the spectrum. Such a gap condition would virtually exclude any eigenvalue resonances.

For finite dimensional state spaces $\mathscr{H}$, the first mathematical proof of the adiabatic theorem was given by BORN and FOCK [16]. These authors considered *simple* eigenvalues with at most finitely many resonances. Further, they assumed that there exists a $\kappa \in \mathbb{N}_0$ such that for each resonance $\omega_\lambda(t_*) = \omega_\mu(t_*)$ a higher order nondegeneracy condition holds,

$$\frac{d^\kappa}{dt^\kappa}(\omega_\lambda - \omega_\mu)\bigg|_{t=t_*} \neq 0,$$

just putting $\kappa = 0$ if there are no resonances at all. By estimating oscillatory integrals as arising in geometric optics, they were able to prove the asymptotic result, [16, Eq. (60)],

$$\langle P_\lambda \psi_\epsilon, \psi_\epsilon \rangle = \langle P_\lambda(0)\psi_*, \psi_* \rangle + O\left(\epsilon^{-\kappa-1}\right).$$

This shows in particular that the rate of convergence in Theorem 1 can be arbitrary slow as a power of the singular perturbation parameter ϵ.

This early work set up the central ideas which, more or less, were followed by all later work: a suitable coordinate change, called "rotating axis representation" in the textbook of MESSIAH [69, §XIII.12], and a simplified motion for comparison purposes, called "adiabatic evolution" by KATO [50].

The work of BORN and FOCK was later extended to infinite dimensional state spaces $\mathcal{H}$ by KATO [50], who, however, restricted himself to the case that the essential spectrum is not excited initially, i.e., $Q(0)\psi_* = 0$. In unpublished work, FRIEDRICHS [32] removed this restriction. Both, KATO and FRIEDRICHS had to use a "gap-condition" for their proof to work.

In a second part [33] of his work, concerned with two-dimensional state spaces, FRIEDRICHS studied the resonant case, restating the result of BORN and FOCK for the case $\kappa = 1$.

The most complete account of the method of "adiabatic evolution" can be found in the more recent work of AVRON, SEILER, and YAFFE [7][8]. Notice that the "gap-condition" [7, cond. (iii), p. 37] is crucial there. Moreover, these authors only consider time-dependent Hamiltonians with a time-*independent* domain of definition. Thus, their result does *not* apply to the two-body Hamiltonians which have been studied in §1.5.

The completely different strategy of proof used in §1.2 was developed by the present author and presented by BORNEMANN and SCHÜTTE [19] for the case of finite dimensional state spaces. Using density operators, this method is physically intuitive and allows to study resonances while being reasonably simple. In the opinion of the present author, this simplicity can also be appreciated in the infinite dimensional case, as soon as one is willing to accept the tools from functional analysis presented in Appendix B.

§2.   Quantum-Classical Coupling: The Infinite Dimensional Case

Continuing our study of the coupling model

$$\ddot{y}_\epsilon^j = -\langle \partial_j H(y_\epsilon)\psi_\epsilon, \psi_\epsilon \rangle, \qquad j = 1, \dots, n,$$

$$i\epsilon \dot{\psi}_\epsilon = H(y_\epsilon)\psi_\epsilon,$$

we observe the following: The classical position y influences the Hamiltonian $H(y)$ slowly compared to the time scale $O(\epsilon)$ of rapid fluctuations in the wave function ψ_ϵ, in fact "infinitely slowly" in the singular limit $\epsilon \to 0$.

Given that only a finite number of isolated eigenvalues (energy levels)[61] $\omega_\lambda(y)$ of the Hamiltonian $H(y)$ are excited initially, we thus would expect, in analogy to the quantum adiabatic theorem, Theorem 1, the *adiabatic invariance* of the energy level probabilities,

$$\langle P_\lambda(y_\epsilon)\psi_\epsilon, \psi_\epsilon \rangle \rightarrow \theta_*^\lambda = \text{const}$$

as $\epsilon \rightarrow 0$. Here, $P_\lambda(y)$ denotes a projection into the eigenspace corresponding to $\omega_\lambda(y)$. This motivates the convergence of the potential energy,

$$\langle H(y_\epsilon)\psi_\epsilon, \psi_\epsilon \rangle \rightarrow U_{\text{BO}}(y_0) = \sum_\lambda \theta_*^\lambda \omega_\lambda(y_0),$$

leading us to the limit equation as

$$\ddot{y}_0^j = -\partial_j U_{\text{BO}}(y_0),$$

which is the well-known *time-dependent Born-Oppenheimer approximation* of quantum theory, cf. [22][41]. Notice that we extend the notion of the Born-Oppenheimer approximation to a finite number of excited energy levels, not just the ground state.

§2.1.   The Singular Limit

We will state and prove the convergence of the coupling model to the Born-Oppenheimer approximation under hypotheses (Q1)-(Q3) below. These hypotheses are more restrictive than (H1)-(H3) in the preceding §1. This is because first, there seems to be no such general existence result available in the literature for nonlinearly coupled Schrödinger equations as for just time-dependent ones, and second, we have to show more than the adiabatic invariance, namely the weak* convergence to the Born-Oppenheimer force of the force term in the first, Newtonian equation of the coupling model.

(Q1) The semi-bounded selfadjoint operator H_0 has the form domain $\mathscr{H}_+$ yielding the rigging $\mathscr{H}_+ \hookrightarrow \mathscr{H} \hookrightarrow \mathscr{H}_-$. The coercivity estimate of the corresponding form is explicitly given by

$$\langle H_0\psi, \psi \rangle \geq \gamma\|\psi\|_{\mathscr{H}_+}^2 - \kappa_1\|\psi\|^2, \qquad \psi \in \mathscr{H}_+,$$

for some $\gamma > 0$ and $\kappa_1 \in \mathbb{R}$.

(Q2) The interaction potential V fulfills $V \in C^2(\overline{\mathbb{R}^n}, \mathscr{B}(\mathscr{H}))$. The uniform bound for V is explicitly given by

$$\|V(y)\|_{\mathscr{B}(\mathscr{H})} \leq \kappa_2, \qquad y \in \mathbb{R}^n,$$

for some $\kappa_2 \in \mathbb{R}$.

---

[61] Recall Footnote 58 on p. 92.

Now, the parameter-dependent Hamiltonian of the system under consideration is

$$H(y) = H_0 + V(y), \qquad y \in \mathbb{R}^n.$$

Since $V(y)$ is a bounded operator, the Kato-Rellich theorem, [76, Theorem X.12], shows that $H(y)$ is a semi-bounded selfadjoint operator with domain

$$D(H(y)) = D(H_0), \qquad y \in \mathbb{R}^n.$$

The corresponding form $h(y; \cdot, \cdot)$ on $\mathscr{H}_+$, defined by

$$h(y; \psi, \phi) = \langle H_0\psi, \phi \rangle + \langle V(y)\psi, \phi \rangle, \qquad \psi, \phi \in \mathscr{H}_+,$$

is uniformly $\mathscr{H}_+$-coercive,

$$h(y; \psi, \psi) \geq \gamma \|\psi\|^2_{\mathscr{H}_+} - (\kappa_1 + \kappa_2)\|\psi\|^2, \qquad y \in \mathbb{R}^n, \ \psi \in \mathscr{H}_+. \qquad \text{(IV.8)}$$

Thus, $H(y)$ extends to a bounded operator $\mathscr{H}_+ \to \mathscr{H}_-$ such that

$$H \in C^2(\mathbb{R}^n, \mathscr{B}(\mathscr{H}_+, \mathscr{H}_-)), \quad \partial_j H = \partial_j V \in C^1(\mathbb{R}^n, \mathscr{B}(\mathscr{H})), \quad j = 1, \ldots, n.$$

The next hypothesis specifies the spectral information we need.

(Q3) There is a number $s \in \mathbb{N}$ and a family $\omega_\lambda \in C^2(\mathbb{R}^n)$ of eigenvalues

$$\omega_\lambda(y) \in \sigma_{\text{disc}}(H(y)), \qquad \lambda = 1, \ldots, s.$$

Further, there is a family $P_\lambda \in C^1(\mathbb{R}^n, \mathscr{K}(\mathscr{H}))$ of orthogonal projections obeying

$$\text{range}\, P_\lambda(y) \subset D(H_0), \qquad H(y)P_\lambda(y) = \omega_\lambda(y)P_\lambda(y), \qquad y \in \mathbb{R}^n.$$

The orthogonal projection $Q = I - P_1 - \ldots - P_s$ is the spectral projection belonging to the remaining part of the spectrum, $\sigma(H(y)) \setminus \{E_1(y), \ldots, E_s(y)\}$, for all $y \in \mathbb{R}^n$.

Now, Definition II.4 translates to the present setting as follows.

Definition 1. Introducing the constants

$$\theta^\lambda_* = \langle P_\lambda(y_*)\psi_*, \psi_* \rangle, \qquad \lambda = 1, \ldots, s,$$

we set

$$U_{\text{BO}}(y) = \sum_{\lambda=1}^{s} \theta^\lambda_* \omega_\lambda(y), \qquad y \in \mathbb{R}^n.$$

The potential U_{BO} will be called the *Born-Oppenheimer potential* belonging to the given initial values $y_* \in \mathbb{R}^n$ and $\psi_* \in D(H_0)$.[62]

---

[62]Notice that $Q(y_*)\psi_* = 0$ already implies $\psi_* \in D(H_0)$.

Theorem 3. *Let hypotheses (Q1), (Q2), and (Q3) be valid. For a sequence $\epsilon \to 0$, and a finite time T, there exist unique sequences $y_\epsilon \in C^2([0,T], \mathbb{R}^n)$ and $\psi_\epsilon \in C^1([0,T], \mathscr{H})$ of solutions of the coupling equations*

$$
\begin{aligned}
\ddot{y}_\epsilon^j &= -\langle \partial_j H(y_\epsilon) \psi_\epsilon, \psi_\epsilon \rangle, \qquad j = 1, \ldots, n, \\
i\epsilon \dot{\psi}_\epsilon &= H(y_\epsilon) \psi_\epsilon,
\end{aligned}
\tag{IV.9}
$$

with initial values $y_\epsilon(0) = y_*$, $\dot{y}_\epsilon(0) = v_*$, and $\psi_\epsilon(0) = \psi_*$ with $\|\psi_*\| = 1$ and $Q(y_*)\psi_* = 0$.

Let U_{BO} be the Born-Oppenheimer potential belonging to the given initial values and $y_{BO} \in C^2([0,T], \mathbb{R}^n)$ the unique solution of the Born-Oppenheimer equation

$$
\ddot{y}_{BO}^j = -\partial_j U_{BO}(y_{BO}), \qquad j = 1, \ldots, n,
$$

with initial values $y_{BO} = y_*$, $\dot{y}_{BO}(0) = v_*$.

If y_{BO} is non-flatly resonant up to order two,[63] the positions and velocities of y_ϵ converge uniformly to those of y_{BO}, i.e., $y_\epsilon \to y_{BO}$ in $C^1([0,T], \mathbb{R}^n)$.

§2.2. The Proof

Again, we follow the four steps of the scheme set up in the illustrative example of §I.2. However, the first three steps are only modifications of the proof given in §1.2 for the adiabatic theorem of quantum mechanics, Theorem 1. Throughout this section the hypotheses of Theorem 3 shall be valid.

Step 1: Energy-Principle. We start with a local existence and uniqueness result.

Lemma 7. *Let be $Q(y_*)\psi_* = 0$. Then, for each $\epsilon > 0$ there is a maximal time $0 < T_\epsilon \le \infty$ such that the coupling model, Eq. (IV.9), has a unique solution $y_\epsilon \in C^2([0, T_\epsilon[, \mathbb{R}^n)$ and $\psi_\epsilon \in C^1([0, T_\epsilon[, \mathscr{H})$. The wave function has the additional regularity*

$$
\psi_\epsilon(t) \in D(H_0), \qquad t \in [0, T_\epsilon[.
$$

If $T_\epsilon < \infty$, then there would be a blow-up in the sense that

$$
\lim_{t \to T_\epsilon} (|y_\epsilon(t)| + |\dot{y}_\epsilon(t)| + \|\psi_\epsilon(t)\|) = \infty.
$$

Proof. We write Eq. (IV.9) as a first order system

$$
\dot{U}_\epsilon = A U_\epsilon + F_\epsilon(U_\epsilon)
$$

---

[63] Recall Definition II.3 for this notion.

on the Hilbert space $\mathscr{X} = \mathbb{R}^n \times \mathbb{R}^n \times \mathscr{H}$. Here,

$$U_\epsilon = \begin{pmatrix} y_\epsilon \\ v_\epsilon \\ \psi_\epsilon \end{pmatrix}, \qquad A = \begin{pmatrix} 0 & I & 0 \\ -I & 0 & 0 \\ 0 & 0 & -i\epsilon^{-1}H_0 \end{pmatrix},$$

and

$$F_\epsilon(U_\epsilon) = \begin{pmatrix} 0 \\ y_\epsilon - \langle \nabla V(y_\epsilon)\psi_\epsilon, \psi_\epsilon \rangle \\ -i\epsilon^{-1}V(y_\epsilon)\psi_\epsilon \end{pmatrix}.$$

The linear operator A is skewadjoint with domain of definition $D(A) = \mathbb{R}^n \times \mathbb{R}^n \times D(H_0)$. By Stone's theorem it generates a C_0-group of unitary operators. The assumptions on the initial data imply that $U_\epsilon(0) \in D(A)$. By hypothesis (Q2), the mapping $F : \mathscr{X} \to \mathscr{X}$ is continuously differentiable. Therefore, the theory of Lipschitz perturbations of linear evolution equations, cf. [90][73, §6.1], is applicable and yields the result.   □

According to the study of the finite dimensional case in §III.3, the energy of the coupling model, Eq. (IV.9), is given as the following time-dependent function:

$$E_\epsilon = \tfrac{1}{2}|\dot y_\epsilon|^2 + \langle H(y_\epsilon)\psi_\epsilon, \psi_\epsilon \rangle.$$

Conservation of energy and norm leads to the following basic estimates.

Lemma 8. *For all $\epsilon > 0$ there is $T_\epsilon = \infty$. One obtains conservation of norm and energy,*

$$\|\psi_\epsilon(t)\| = \|\psi_*\| = 1, \qquad E_\epsilon(t) = E_* = \tfrac{1}{2}|v_*|^2 + \langle H(y_*)\psi_*, \psi_* \rangle,$$

for all $t \geq 0$, and, for a finite time interval $[0,T]$, the uniform bounds

$$y_\epsilon = O(1) \quad in \quad C^2([0,T], \mathbb{R}^n), \qquad \psi_\epsilon = O(1) \quad in \quad L^\infty([0,T], \mathscr{H}_+),$$

as $\epsilon \to 0$.

Proof. Using the Schrödinger equation for ψ_ϵ we obtain that

$$\frac{d}{dt}\langle \psi_\epsilon, \psi_\epsilon \rangle = i\epsilon^{-1}\langle \psi_\epsilon, H(y_\epsilon)\psi_\epsilon \rangle - i\epsilon^{-1}\langle H(y_\epsilon)\psi_\epsilon, \psi_\epsilon \rangle = 0,$$

i.e., the conservation of norm. Since $\psi_\epsilon(t) \in D(H_0) = D(H(y_\epsilon(t)))$, virtually the same argument which led in the proof of Lemma 1 to Eq. (IV.1)[64] gives here that

$$\frac{d}{dt}\langle H(y_\epsilon)\psi_\epsilon, \psi_\epsilon \rangle = \left\langle \left(\frac{d}{dt}H(y_\epsilon)\right)\psi_\epsilon, \psi_\epsilon \right\rangle = \sum_{j=1}^n \langle \partial_j H(y_\epsilon)\psi_\epsilon, \psi_\epsilon \rangle \dot y_\epsilon^j.$$

---

[64]The "Hellmann-Feynman theorem."

This shows that the time-derivative of the energy vanishes,

$$\dot{E}_\epsilon = \sum_j \left(\ddot{y}_\epsilon^j + \langle \partial_j H(y_\epsilon)\psi_\epsilon, \psi_\epsilon \rangle \right) \cdot \dot{y}_\epsilon^j = 0.$$

Thus, for any time t of existence, the uniform coercivity, Eq. (IV.8), implies the following bound:

$$\gamma \|\psi_\epsilon(t)\|_{\mathscr{H}_+}^2 \leq \langle H(y_\epsilon(t))\psi_\epsilon(t), \psi_\epsilon(t) \rangle + (\kappa_1 + \kappa_2)\|\psi_\epsilon(t)\|^2 \leq E_* + (\kappa_1 + \kappa_2).$$

Likewise, one gets the uniform bound

$$\tfrac{1}{2}|\dot{y}_\epsilon(t)|^2 \leq E_* - \langle H(y_\epsilon(t))\psi_\epsilon(t), \psi_\epsilon(t) \rangle \leq E_* + (\kappa_1 + \kappa_2).$$

Integration yields a bound for y_ϵ which grows only *linearly* in time. This way, there is no blow-up in the sense of Lemma 7 and therefore $T_\epsilon = \infty$. Finally, by the bounds proven above, the force term $\langle \partial_j H(y_\epsilon)\psi_\epsilon, \psi_\epsilon \rangle$ is uniformly bounded yielding the uniform bound for $\ddot{y}_\epsilon$.   □

As in the proof of Theorem 1 we introduce the time-dependent density operator

$$\rho_\epsilon = \langle \cdot, \psi_\epsilon \rangle \psi_\epsilon \;\in\; L^\infty([0,T], \mathscr{B}(\mathscr{H}_-, \mathscr{H}_+)) \cap C([0,T], \mathscr{J}_1(\mathscr{H})).$$

The uniform bounds of Lemma 8 directly imply the following lemma.

Lemma 9. *There is a subsequence of ϵ, denoted by ϵ again, and functions $y_0 \in C^{1,1}([0,T], \mathbb{R}^n)$ and $\rho_0 \in L^\infty([0,T], \mathscr{B}(\mathscr{H}_-, \mathscr{H}_+))$ such that:*

$$
\begin{array}{llll}
\text{(i)} & y_\epsilon \to y_0 & \text{in} & C^1([0,T], \mathbb{R}^n), \\
\text{(ii)} & \ddot{y}_\epsilon \overset{*}{\rightharpoonup} \ddot{y}_0 & \text{in} & L^\infty([0,T], \mathbb{R}^n), \\
\text{(iii)} & \rho_\epsilon \overset{*}{\rightharpoonup} \rho_0 & \text{in} & L^\infty([0,T], \mathscr{J}_1(\mathscr{H})), \\
\text{(iv)} & \rho_\epsilon \sigma^* \overset{*}{\rightharpoonup} \rho_0 \sigma^* & \text{in} & L^\infty([0,T], \mathscr{J}_1(\mathscr{H}_+)), \\
\text{(v)} & \sigma^* \rho_\epsilon \overset{*}{\rightharpoonup} \sigma^* \rho_0 & \text{in} & L^\infty([0,T], \mathscr{J}_1(\mathscr{H}_-)), \\
\text{(vi)} & \rho_\epsilon H(y_\epsilon) \overset{*}{\rightharpoonup} \rho_0 H(y_0) & \text{in} & L^\infty([0,T], \mathscr{J}_1(\mathscr{H}_+)), \\
\text{(vii)} & H(y_\epsilon)\rho_\epsilon \overset{*}{\rightharpoonup} H(y_0)\rho_0 & \text{in} & L^\infty([0,T], \mathscr{J}_1(\mathscr{H}_-)).
\end{array}
$$

Proof. The extended Arzelà-Ascoli theorem, Principle I.4, yields the assertion about y_ϵ. Virtually the same proof as of Lemma 2 gives the assertions about ρ_ϵ. The only point to mention is that

$$\sigma H(y_\epsilon) \to \sigma H(y_0) \quad \text{in} \; L^\infty([0,T], \mathscr{J}_1(\mathscr{H}_+))$$

and

$$H(y_\epsilon)\sigma \to H(y_0)\sigma \quad \text{in} \; L^\infty([0,T], \mathscr{J}_1(\mathscr{H}_-))$$

because of the uniform convergence $y_\epsilon \to y_0$.   □

Step 2: The Weak Virial Theorem.   Also, the analogue of Lemma 3 holds true with literally the same proof.

Lemma 10. (Weak Virial Theorem). *There holds the commutativity relation*

$$\rho_0(t)\, H(y_0(t)) \subset H(y_0(t))\, \rho_0(t)$$

for almost all $t \in [0, T]$ *as unbounded, densely defined operators in* $\mathscr{H}$.

As an immediate corollary we obtain the following analogue of Lemma 4, again with literally the same proof.

Lemma 11. *Let there be essentially no resonances of order two along the limit* y_0. *Then, for almost all* $t \in [0, T]$, *one gets the following commutativity relations in* $\mathscr{B}(\mathscr{H})$:

$$[\rho_0(t), Q(y_0(t))] = 0, \qquad [\rho_0(t), P_\lambda(y_0(t))] = 0, \quad \lambda = 1, \ldots, s.$$

Step 3: Adiabatic Invariance of the Action.   We introduce the energy level probabilities (actions)

$$\theta_\epsilon^\lambda = \langle P_\lambda(y_\epsilon)\psi_\epsilon, \psi_\epsilon \rangle, \qquad \theta_\epsilon^Q = \langle Q(y_\epsilon)\psi_\epsilon, \psi_\epsilon \rangle.$$

The adiabatic invariance of these quantities follows directly from the commutativity relation Lemma 11.

Lemma 12. *Let there be essentially no resonances of order two along the limit* y_0. *Then, there are the uniform convergences*

$$\theta_\epsilon^\lambda \to \theta_*^\lambda = \langle P_\lambda(y_*)\psi_*, \psi_* \rangle, \qquad \theta_\epsilon^Q \to \theta_*^Q = \langle Q(y_*)\psi_*, \psi_* \rangle,$$

in the space $C[0, T]$. *If* $Q(y_*)\psi_* = 0$, *then*

$$\theta_*^Q = 0, \qquad \rho_0 Q(y_0) = Q(y_0)\rho_0 Q(y_0) = 0.$$

Proof. The proof is virtually the same as of Lemma 5 and Corollaries 1 and 3. We should only mention that one is using the strong convergence

$$P_\lambda(y_\epsilon) \to P_\lambda(y_0) \quad \text{in} \quad C^1([0, T], \mathscr{K}(\mathscr{H})),$$

which follows from hypothesis (Q3) and $y_\epsilon \to y_0$ in $C^1([0, T], \mathbb{R}^n)$.   □

We need the following generalization of Corollary 3.

Lemma 13. *Let there be essentially no resonances of order two along the limit* y_0. *If* $Q(y_*)\psi_* = 0$, *one gets for a sequence* $A_\epsilon \to A_0$, *strongly converging in* $L^\infty([0, T], \mathscr{B}(\mathscr{H}))$, *the weak* convergence*

$$\operatorname{tr}(A_\epsilon \rho_\epsilon) \xrightarrow{*} \operatorname{tr}(A_0 \rho_0) \quad \text{in} \quad L^\infty([0, T], \mathbb{C}).$$

In particular, one obtains tr $\rho_0 = 1$ in $L^\infty[0,T]$ and, for $j = 1,\ldots,n$, the abstract limit equation

$$\ddot{y}_0^j = -\operatorname{tr}(\partial_j H(y_0) \cdot \rho_0) \quad \text{in} \quad L^\infty[0,T]. \tag{IV.10}$$

Proof. The definition of Q yields

$$\operatorname{tr}(A_\epsilon \rho_\epsilon) = \operatorname{tr}(Q(y_\epsilon)A_\epsilon \rho_\epsilon) + \sum_\lambda \operatorname{tr}(P_\lambda(y_\epsilon)A_\epsilon \rho_\epsilon). \tag{*}$$

For the compact operators form an ideal in the algebra of bounded operators, we obtain that $P_\lambda(y_\epsilon)A_\epsilon \to P_\lambda(y_0)A_0$ in $L^\infty([0,T], \mathcal{K}(\mathcal{H}))$. Thus, by Lemma B.5,

$$\operatorname{tr}(P_\lambda(y_\epsilon)A_\epsilon \rho_\epsilon) \overset{*}{\rightharpoonup} \operatorname{tr}(P_\lambda(y_0)A_0\rho_0) \quad \text{in} \quad L^\infty([0,T],\mathbb{C}).$$

On the other hand, since the norms of the sequence A_ϵ in $L^\infty([0,T], \mathcal{B}(\mathcal{H}))$ are uniformly bounded by some constant K, we obtain the estimate

$$|\operatorname{tr}(Q(y_\epsilon)A_\epsilon \rho_\epsilon)| \le K \cdot \|Q(y_\epsilon)\psi_\epsilon\|.$$

Since

$$\|Q(y_\epsilon)\psi_\epsilon\|^2 = \langle Q(y_\epsilon)\psi_\epsilon, \psi_\epsilon \rangle = \theta_\epsilon^Q \to 0 \quad \text{in} \quad C[0,T],$$

we finally get, using $\rho_0 Q(y_0) = 0$,

$$\operatorname{tr}(Q(y_\epsilon)A_\epsilon \rho_\epsilon) \to 0 = \operatorname{tr}(Q(y_0)A_0\rho_0) \quad \text{in} \quad L^\infty([0,T],\mathbb{C}).$$

The sum in (*) being *finite* we have thus shown the asserted weak* convergence of the trace expression.

The choice $A_\epsilon = I$ leads to $1 = \operatorname{tr}\rho_\epsilon \overset{*}{\rightharpoonup} \operatorname{tr}\rho_0$, i.e., $\operatorname{tr}\rho_0 = 1$.

Hypothesis (Q2) makes the choice $A_\epsilon = \partial_j H(y_\epsilon)$ admissible; showing the weak convergence of the force term in the first, Newtonian equation of the coupling model, Eq. (IV.9),

$$\langle \partial_j H(y_\epsilon)\psi_\epsilon, \psi_\epsilon \rangle = \operatorname{tr}(\partial_j H(y_\epsilon)\rho_\epsilon) \overset{*}{\rightharpoonup} \operatorname{tr}(\partial_j H(y_0)\rho_0) \quad \text{in} \quad L^\infty[0,T].$$

This, together with the weak* convergence $\ddot{y}_\epsilon \overset{*}{\rightharpoonup} \ddot{y}_0$, yields the asserted *abstract limit equation.* □

Notice that in this Lemma the values of the sequence A_ϵ are *not* restricted to the compact operators. This is in sharp contrast to the general result of Lemma B.5, and only possible because of the specific structure of the sequence ρ_ϵ implied by the hypothesis $Q(y_*)\psi_* = 0$.

Step 4: Identification of the Limit Mechanical System. First, we show that the force term of the abstract limit equation (IV.10) is given by the Born-Oppenheimer potential.

Lemma 14. *Let there be essentially no resonances of order two along the limit y_0. Then, there holds*

$$\mathrm{tr}(\partial_j H(y_0) \cdot \rho_0) = \partial_j U_{\mathrm{BO}}(y_0).$$

Proof. Differentiating

$$H(y)P_\lambda(y) = \omega_\lambda(y)P_\lambda(y)$$

with respect to y^j yields, by the closedness of $H(y)$ and hypothesis (Q2), that

$$\partial_j H \cdot P_\lambda + H \cdot \partial_j P_\lambda \cdot P_\lambda = \partial_j \omega_\lambda \cdot P_\lambda + \omega_\lambda \cdot \partial_j P_\lambda \cdot P_\lambda. \qquad (*)$$

Hence, recalling that $\partial_j H(y) \in \mathscr{B}(\mathscr{H})$, we obtain for $\psi \in \mathscr{H}$ that

$$\partial_j P_\lambda(y) \cdot P_\lambda(y)\psi \in D(H_0).$$

Spectral theory of unbounded selfadjoint operators shows that[65]

$$P_\lambda(y_0)H(y_0) \subset H(y_0)P_\lambda(y_0) = \omega_\lambda(y_0)P_\lambda(y_0)$$

for almost all $t \in [0,T]$. Thus, evaluating $(*)$ at $y = y_0$ and multiplying with $P_\lambda(y_0)$ from the left yields

$$P_\lambda(y_0) \cdot \partial_j H(y_0) \cdot P_\lambda(y_0) = \partial_j \omega_\lambda(y_0) \cdot P_\lambda(y_0).$$

This way, by well-known properties of the trace and by the commutativity result Lemma 11, we get

$$
\begin{aligned}
\mathrm{tr}(\partial_j H(y_0)\rho_0 \cdot P_\lambda(y_0)) &= \mathrm{tr}(P_\lambda(y_0)\partial_j H(y_0)P_\lambda(y_0) \cdot \rho_0) \\
&= \partial_j \omega_\lambda(y_0) \cdot \mathrm{tr}(P_\lambda(y_0)\rho_0) \\
&= \theta_*^\lambda \cdot \partial_j \omega_\lambda(y_0).
\end{aligned}
$$

By Lemma 12 we have $\rho_0 Q(y_0) = 0$ which finally shows that

$$
\begin{aligned}
\mathrm{tr}(\partial_j H(y_0)\rho_0) &= \mathrm{tr}(\partial_j H(y_0)\rho_0 \cdot Q(y_0)) + \sum_\lambda \mathrm{tr}(\partial_j H(y_0)\rho_0 \cdot P_\lambda(y_0)) \\
&= \sum_\lambda \theta_*^\lambda \cdot \partial_j \omega_\lambda(y_0) = \partial_j U_{\mathrm{BO}}(y_0).
\end{aligned}
$$

Recall that the limit energy level probabilities θ_*^λ are constant in time. $\square$

Now, to summarize, we have shown that *if* there were essentially no resonances of order two along the up to now *inaccessible* limit y_0, the equality $y_0 = y_{\mathrm{BO}}$ would hold. As in §II.2.4, one can decide on this equality by looking at the resonance properties of the *accessible* function y_{BO}.

---

[65]E.g., the particular result [83, Eq. (16), p. 345].

Lemma 15. *If y_{BO} is non-flatly resonant up to order two, there are only finitely many resonances of order two y_0, and one gets*

$$y_0 = y_{BO}.$$

Proof. One proceeds virtually along the same argument as in the proof of Lemma II.16. The only difference is that it suffices to consider resonances of order *two*.   □

From this lemma we conclude that the limit $y_0 = y_{BO}$ is independent of the subsequence chosen. Thus, by Principle I.5, we may discard all extractions of subsequences and have proven the uniform convergence $y_\epsilon \to y_{BO}$, as was asserted in the statement of Theorem 3.

§2.3.   An Example

Illustrating the relation of the quantum-classical coupling model to a full quantum model, BORNEMANN, NETTESHEIM, and SCHÜTTE [17] studied the simple example of a collinear collision of a "classical" particle with a quantum harmonic oscillator. Actually, this example fits into the framework of Theorem 3.

The Hamiltonian of the one-dimensional harmonic oscillator is given by

$$H_0 = -\frac{\partial^2}{\partial x^2} + c x^2, \qquad c > 0,$$

on the Hilbert space $\mathscr{H} = L^2(\mathbb{R})$. The interaction potential with the classical particle is given by

$$V(y, x) = a \exp(-b|x - y|^2)$$

with some constants $a, b > 0$. Thus hypotheses (Q1) and (Q2) of §2.1 are valid. Theorem XIII.16 of [77] tells us that $H(y) = H_0 + V(y, \cdot)$ has a purely discrete spectrum. Let only finitely many eigenvalues of $H(y_*)$ be excited by the initial wave function ψ_*. Since $H(y)$ depends on the *one-dimensional* parameter y only, these initially excited eigenvalues and the corresponding eigenspaces vary smoothly in y—at least as long as there are no resonances, cf. [51]. Thus hypothesis (Q3) will be valid and Theorem 3 is applicable.

§2.4.   Remarks on the Born-Oppenheimer Approximation

The time-dependent Born-Oppenheimer approximation can be viewed as a (partial) semiclassical approximation of a full quantum description of the underlying two-particle system. HAGEDORN [40][41] has studied this approximation in detail, showing that the order of approximation is given by $O(\epsilon^{1/2})$ for Gaussian initial preparations. Recall that $\epsilon^2 = m/M \ll 1$ denotes the mass ratio of the two particles. This approximation result,

together with our new limit result Theorem 3, gives a further justification
of the quantum-classical coupling model, different from the one given in
the work of BORNEMANN, NETTESHEIM, and SCHÜTTE [17].

For approximating the full quantum model by the Born-Oppenheimer
model, one has to exclude any *crossings*, or *resonances*, of the energy levels involved. Restricting himself to initial excitations of the ground state,
HAGEDORN [42][43] studied *normal forms* of generic energy crossings. As
discussed for the Hessians in §II.1.7, these normal forms are classified
according to the codimension of the corresponding manifold of resonant
Hamiltonians. HAGEDORN has proven that generically there are only 11
distinct types of crossings. Depending on the type, these crossings have
codimension 1,2,3, or 5.

Now, crossings of codimension one correspond to those Born-Oppenheimer solutions which we have called "non-flatly resonant up to order
two." Thus, Theorem 3 is applicable here, showing that the relation of the
Born-Oppenheimer model and the coupling model is not affected by these
crossings.

The only type of a codimension two crossing corresponds to a Hamiltonian which is not "smoothly diagonalizable" in the excited part of the spectrum. Here, the relation between the Born-Oppenheimer and the quantum-
classical coupling model can be affected by *Takens chaos*. An example for
this to happen which is analogous to the one given in §II.4 can be found
in the work of BORNEMANN and SCHÜTTE [19][88]. In the fields where the
coupling model is currently used, the relevance of this effect has yet to be
studied. There are, however, some indications [88] that this effect reflects
non-adiabatic excitations in a full quantum description of the underlying
system.

Appendix A:
Eigenvalue Resonances of Codimension Two

In this part of the appendix we study a parameter-dependent symmetric matrix H at a *generic* eigenvalue resonance.[66] Here, the term "generic" means that the appearance of the resonance is *stable* with respect to the most general class of *perturbations* which preserves the symmetry $H = H^T$. We remind the reader, that a problem can call for a smaller class of perturbations that preserves more structural symmetries. An example can be found in §II.1.7.

The set of matrices that have at least one eigenvalue with multiplicity greater than one has *codimension two* in the set of all real symmetric matrices of a certain size. This "loss" of two dimension can be explained as follows. If we represent a symmetric matrix H by its diagonalization $H = S^T D S$, we understand that one dimension is lost by the eigenvalue resonance in the diagonal matrix D. Another dimension is lost, however, in the orthogonal matrix S since the corresponding eigenspace of dimension two can freely be rotated without changing the resulting matrix H.

A general formula for calculating the codimension of more general resonance patterns can be found in the work of VON NEUMANN and WIGNER [71].

Transversality theory [4] teaches that we need at least a two-parameter dependence of H to obtain a generic, i.e., transversal, crossing with the resonance manifold of codimension two. We can always freeze some of the parameters and, by standard perturbation theory [51], follow smoothly the two-dimensional eigenspace that belongs to the resonance. This is because the resonance is locally separated from the rest of the spectrum.

Thus, it suffices to study a family $H(x)$ of two-by-two matrices that depends smoothly on two parameters, $x = (x^1, x^2) \in \mathbb{R}^2$. Now, suppose there is generically a twofold eigenvalue at $x = 0$. To simplify, we transform to the trace-free matrix

$$H_0(x) = H(x) - \tfrac{1}{2}\operatorname{tr} H(x) \cdot I,$$

leaving the genericity of the eigenvalue resonance untouched. Notice, that the eigenvectors remain unchanged and the eigenvalues are symmetrized

---

[66] "Resonance" stands for a resonance relation of order 2, i.e., for the *equality* of two different eigenvalue families at certain parameter values.

with respect to zero. Thus, H_0 has a twofold eigenvalue $\lambda = 0$ at $x = 0$. We denote the coefficients of H_0 by

$$H_0(x) = \begin{pmatrix} \xi^1(x) & \xi^2(x) \\ \xi^2(x) & -\xi^1(x) \end{pmatrix}, \qquad x = (x^1, x^2).$$

By definition of the term "generic," there is in the space of coefficients of symmetric two-by-two matrices a transversal intersection of the hypersurface $\{(\xi^1(x), -\xi^1(x), \xi^2(x)) : x \in \mathbb{R}^2\}$ and the curve $\{(\lambda, \lambda, 0) : \lambda \in \mathbb{R}\}$ at the parameter values $(x, \lambda) = (0, 0)$. Thus, we have

$$0 \neq \begin{vmatrix} \partial_1 \xi^1(0) & \partial_2 \xi^1(0) & 1 \\ -\partial_1 \xi^1(0) & -\partial_2 \xi^1(0) & 1 \\ \partial_1 \xi^2(0) & \partial_2 \xi^2(0) & 0 \end{vmatrix} = -2 \begin{vmatrix} \partial_1 \xi^1(0) & \partial_2 \xi^1(0) \\ \partial_1 \xi^2(0) & \partial_2 \xi^2(0) \end{vmatrix}.$$

In other words, by referring to the inverse function theorem, the eigenvalue resonance is generic if and only if $x \mapsto \xi(x)$ is a smooth coordinate transformation that maps a neighborhood of $x = 0$ bijectively to a neighborhood of $\xi(0) = 0$.

We finally show that $H_0(x)$ and, a forteriori, $H(x)$ do *not* have a smooth spectral decomposition in a neighborhood of $x = 0$. By what we have seen it suffices to prove this claim in ξ-coordinates.

The eigenvalues of H_0 are $\lambda_1(\xi) = -|\xi|$ and $\lambda_2(\xi) = |\xi|$. Excluding the origin $\xi = 0$ and using polar coordinates,

$$\xi^1 = r \cos \phi, \qquad \xi^2 = r \sin \phi,$$

yields the corresponding eigenvectors as given by

$$e_1(\xi) = \begin{pmatrix} -\sin(\phi/2) \\ \cos(\phi/2) \end{pmatrix}, \qquad e_2(\xi) = \begin{pmatrix} \cos(\phi/2) \\ \sin(\phi/2) \end{pmatrix}.$$

The occurrence of the argument $\phi/2$ shows that these eigenvectors are defined up to a sign only. For a unique and smooth representation we have to cut the ξ-plane along a half-axis, e.g., along $\phi = 3\pi/2$. The eigenvectors cannot, however, be smoothly continued beyond that cut, but instead change their mutual roles there. Summarizing, we have proven the following theorem.[67]

Theorem A.1. *A parameter-dependent family of real symmetric matrices does not have a smooth spectral decomposition in a neighborhood of a generic eigenvalue resonance of codimension two.*

---

[67]The matrix $H_0(\xi)$ is just the famous example of RELLICH [80, §2][51, Chap. 2, Example 5.12] for a smooth symmetric matrix which is not smoothly diagonalizable. The study above shows that this example not only occurs naturally but, in a way, unavoidably.

Appendix B:
Advanced Tools from Functional Analysis

§1. Weak* Convergence of Operator-Valued Functions

This part of the appendix collects all those facts about trace-class-operator-valued functions that are applied to the time-dependent density operators in Chapter IV. In particular, we prove an extraction principle, Theorem B.3, by using general facts about Lebesgue-Bochner spaces. As far as we know, there is no explicit reference for Theorem B.3 in the accessible literature. Though not difficult at all, the material of §B.1.5 seems to be entirely new.

§1.1. Trace Class Operators

Here, we recall the basic facts about compact operators and trace class operators on a Hilbert space $\mathscr{H}$. If not stated otherwise, these facts can be found in the textbooks of REED and SIMON [78, Chap. VI.5–6][76, p. 41ff], or in either one of the monographs of RINGROSE [81], SCHATTEN [85], or SIMON [92].

We consider a *separable*, complex Hilbert space $\mathscr{H}$, the inner product $\langle \cdot, \cdot \rangle$ of which is linear in the first factor and conjugate linear in the second. Three important spaces of linear endomorphisms on $\mathscr{H}$ are

- the space $\mathscr{B}(\mathscr{H})$ of *bounded* operators,

- the space $\mathscr{K}(\mathscr{H})$ of *compact* operators,

- the space $\mathscr{F}(\mathscr{H})$ of operators of *finite rank*.

There are the inclusions $\mathscr{F}(\mathscr{H}) \subset \mathscr{K}(\mathscr{H}) \subset \mathscr{B}(\mathscr{H})$. For any compact operator $A \in \mathscr{K}(\mathscr{H})$ there exists a unique decreasing sequence $\{\lambda_j(A)\}$ of nonnegative real numbers with $\lambda_j(A) \to 0$, and two orthonormal sets $\{\psi_j\}$, $\{\phi_j\}$ in $\mathscr{H}$, such that

$$A = \sum_{j=1}^{\infty} \lambda_j(A) \langle \cdot, \psi_j \rangle \phi_j. \tag{B.1}$$

This sum converges in the operator norm. The numbers $\lambda_j(A)$ are called the *singular values* of A, and Eq. (B.1) the *singular value decomposition* or *canonical form* of A. In particular, we obtain

$$\mathcal{K}(\mathcal{H}) = \overline{\mathcal{F}(\mathcal{H})}^{\mathcal{B}(\mathcal{H})}. \tag{B.2}$$

A compact operator A belongs to the *trace class* $\mathcal{J}_1(\mathcal{H})$ if and only if the singular values $\{\lambda_j(A)\}$ are summable, $\{\lambda_j(A)\} \in \ell_1$. This space $\mathcal{J}_1(\mathcal{H})$ is complete with respect to the *trace class norm*

$$\|A\|_{\mathcal{J}_1} = \sum_{j=1}^{\infty} \lambda_j(A), \qquad A \in \mathcal{J}_1(\mathcal{H}).$$

The operators of finite rank are dense in $\mathcal{J}_1(\mathcal{H})$,

$$\mathcal{J}_1(\mathcal{H}) = \overline{\mathcal{F}(\mathcal{H})}^{\mathcal{J}_1(\mathcal{H})}. \tag{B.3}$$

The trace class constitutes a two-sided operator *ideal* in $\mathcal{B}(\mathcal{H})$: for each $A \in \mathcal{J}_1(\mathcal{H})$ and $B \in \mathcal{B}(\mathcal{H})$ we obtain $AB, BA \in \mathcal{J}_1(\mathcal{H})$ with

$$\|AB\|_{\mathcal{J}_1} \leq \|A\|_{\mathcal{J}_1} \cdot \|B\|, \qquad \|BA\|_{\mathcal{J}_1} \leq \|A\|_{\mathcal{J}_1} \cdot \|B\|. \tag{B.4}$$

The *trace* is defined as a linear form on $\mathcal{J}_1(\mathcal{H})$ by

$$\operatorname{tr} A = \sum_j \langle A\phi_j, \phi_j \rangle, \qquad A \in \mathcal{J}_1(\mathcal{H}),$$

for every orthonormal basis $\{\phi_j\}$ of $\mathcal{H}$. The sum converges absolutely and its value is *independent* of the chosen basis. Important properties of the trace are

$$\operatorname{tr} A^* = \overline{\operatorname{tr} A}, \qquad |\operatorname{tr} A| \leq \operatorname{tr} |A| = \|A\|_{\mathcal{J}_1(\mathcal{H})}, \qquad A \in \mathcal{J}_1(\mathcal{H}), \tag{B.5}$$

and

$$\operatorname{tr}(AB) = \operatorname{tr}(BA), \qquad A \in \mathcal{J}_1(\mathcal{H}), \ B \in \mathcal{B}(\mathcal{H}).$$

For operators of rank one, we trivially get

$$\operatorname{tr}(\langle \cdot, \psi \rangle \phi) = \langle \phi, \psi \rangle, \qquad \|\langle \cdot, \psi \rangle \phi\|_{\mathcal{J}_1} = \|\langle \cdot, \psi \rangle \phi\| = \|\phi\| \cdot \|\psi\|. \tag{B.6}$$

The $\mathcal{J}_1$-norm is obtained from the fact that there is only one nonzero singular value $\lambda_1 = \|\phi\| \cdot \|\psi\|$. The Cauchy-Schwarz inequality yields the expression for the operator norm.

Eqs. (B.4) and (B.5) show in particular that

(i) for every $A \in \mathcal{J}_1(\mathcal{H})$ the map $B \mapsto \operatorname{tr}(AB)$ is a continuous linear form on $\mathcal{K}(\mathcal{H})$,

(ii) for every $B \in \mathscr{B}(\mathscr{H})$ the map $A \mapsto \mathrm{tr}(AB)$ is a continuous linear form on $\mathscr{J}_1(\mathscr{H})$.

In fact, all elements of the dual spaces $\mathscr{K}^*(\mathscr{H})$ and $\mathscr{J}_1^*(\mathscr{H})$ are obtained this way: the above mappings induce *natural* isomorphisms

$$\mathscr{K}^*(\mathscr{H}) = \mathscr{J}_1(\mathscr{H}), \qquad \mathscr{J}_1^*(\mathscr{H}) = \mathscr{B}(\mathscr{H}). \qquad (B.7)$$

Notice, however, that there exists *no* predual space of $\mathscr{K}(\mathscr{H})$ for infinite dimensional $\mathscr{H}$.

We close the collection of facts about compact operators and trace class operators with an easy consequence of the density properties (B.2) and (B.3).

Lemma B.1. *The Banach spaces* $\mathscr{K}(\mathscr{H})$ *and* $\mathscr{J}_1(\mathscr{H})$ *are separable.*

Proof. Let $\{\chi_j\}$ be a countable dense set of $\mathscr{H}$. We define the separable space

$$\mathscr{F}_S = \mathrm{span}\{\langle \cdot, \chi_i \rangle \chi_j : i, j \in \mathbb{N}\}.$$

The norm formula (B.6) shows that every rank one operator, and therefore every operator of finite rank, can be approximated arbitrarily well by elements of $\mathscr{F}_S$—both with respect to the operator norm and the $\mathscr{J}_1$-norm. Thus, using (B.2) and (B.3), we obtain

$$\mathscr{K}(\mathscr{H}) = \overline{\mathscr{F}_S}^{\mathscr{B}(\mathscr{H})}, \qquad \mathscr{J}_1(\mathscr{H}) = \overline{\mathscr{F}_S}^{\mathscr{J}_1(\mathscr{H})},$$

which proves the assertions.   □

§1.2. Lebesgue-Bochner Spaces

Let $\Omega \subset \mathbb{R}^d$ be a bounded open set endowed with the Lebesgue measure. If $\mathscr{X}$ is a Banach space and $1 \leq p \leq \infty$, then $L^p(\Omega, \mathscr{X})$ stands for the *Lebesgue-Bochner space* of p-summable functions on Ω which have values in $\mathscr{X}$. A definition of these spaces can be found in DIESTEL'S and UHL'S survey on *vector measures*, [24, Chap. IV, §1]. We recall two important facts.

Theorem B.1. *If* $\mathscr{X}$ *is separable, the space* $L^1(\Omega, \mathscr{X})$ *is separable.*

Proof. According to [98, §11, Theorem 4], the Lebesgue measure on Ω is separable. Therefore, one can construct a *countable* set of simple functions which is dense in $L^1(\Omega, \mathscr{X})$. The proof of this fact is literally the same as of [98, Chap. 4, §20, Theorem 1], with the only exception that one has to take as the set of coefficients a countable dense subset of $\mathscr{X}$ instead of the rational complex numbers.   □

Theorem B.2. *If the dual space $\mathscr{X}^*$ is separable, there is a natural iso-morphism*

$$L^1(\Omega, \mathscr{X})^* = L^\infty(\Omega, \mathscr{X}^*).$$

Proof. The Dunford-Pettis theorem, [24, Chap. III, §3, Theorem 1], states that a separable dual space $\mathscr{X}^*$ has the *Radon-Nikodým property*. A theorem of Bochner and Taylor, [24, Chap. IV, §1, Theorem 1], shows that $L^1(\Omega, \mathscr{X})^* = L^\infty(\Omega, \mathscr{X}^*)$ if and only if $\mathscr{X}^*$ has the Radon-Nikodým property. □

§1.3.   Spaces of Trace-Class-Operator-Valued Functions

There is an extraction principle for the space $L^\infty(\Omega, \mathscr{J}_1(\mathscr{H}))$ of trace-class-operator-valued functions which is completely analogous to the extraction Principle I.3 for $L^\infty(\Omega, \mathbb{C})$. We continue to use the notation of §§B.1.1 and B.1.2.

Theorem B.3. *There is a natural isomorphism*

$$L^1(\Omega, \mathscr{K}(\mathscr{H}))^* = L^\infty(\Omega, \mathscr{J}_1(\mathscr{H})).$$

In particular, we obtain that a sequence $\{X_\epsilon\}$ of $L^\infty(\Omega, \mathscr{J}_1(\mathscr{H}))$ converges weakly* to $X_0 \in L^\infty(\Omega, \mathscr{J}_1(\mathscr{H}))$, $X_\epsilon \overset{*}{\rightharpoonup} X_0$, if and only if

$$\int_\Omega \operatorname{tr}(X_\epsilon(t) \cdot Y(t))\, dt \to \int_\Omega \operatorname{tr}(X_0(t) \cdot Y(t))\, dt, \qquad Y \in L^1(\Omega, \mathscr{K}(\mathscr{H})).$$

Let $\{X_\epsilon\}$ be a bounded sequence in the space $L^\infty(\Omega, \mathscr{J}_1(\mathscr{H}))$. Then, there is a subsequence $\{\epsilon'\}$ and a function $X_0 \in L^\infty(\Omega, \mathscr{J}_1(\mathscr{H}))$, such that $X_{\epsilon'} \overset{*}{\rightharpoonup} X_0$. Principle I.5 is applicable to the sequence $\{X_\epsilon\}$.

Proof. Lemma B.1 and Theorems B.1–B.2, as well as the natural isomorphism (B.7), show that, first, the space $L^1(\Omega, \mathscr{K}(\mathscr{H}))$ is separable and, second, there is a natural isomorphism $L^1(\Omega, \mathscr{K}(\mathscr{H}))^* = L^\infty(\Omega, \mathscr{J}_1(\mathscr{H}))$. This natural isomorphism leads directly to the given characterization of the weak*-convergence in $L^\infty(\Omega, \mathscr{J}_1(\mathscr{H}))$.

The Alaoglu theorem [83, Theorem 3.15] states that a closed ball in $L^\infty(\Omega, \mathscr{J}_1(\mathscr{H}))$ is compact with respect to the weak*-topology. Since the predual space $L^1(\Omega, \mathscr{K}(\mathscr{H}))$ is separable, the weak*-topology is *metrizable* on closed balls [83, Theorem 3.16]. Hence, bounded sequences have weak*-convergent subsequences and Principle I.5 is applicable. □

The reader should notice that this theorem is *not* a trivial corollary from the fact that $\mathscr{X}^*(\mathscr{H}) = \mathscr{J}_1(\mathscr{H})$. For instance, the space $\mathscr{B}(\mathscr{H})$, belonging to an *infinite* dimensional Hilbert space $\mathscr{H}$, lacks the Radon-Nikodým property, cf. [24, Chap. VII, §7], and therefore we have

$$L^1(\Omega, \mathscr{J}_1(\mathscr{H}))^* \neq L^\infty(\Omega, \mathscr{B}(\mathscr{H})),$$

despite the fact that $\mathscr{J}_1^*(\mathscr{H}) = \mathscr{B}(\mathscr{H})$.

There is a simple but extremely useful criterion for testing weak* convergence in $L^\infty(\Omega, \mathscr{J}_1(\mathscr{H}))$.

Lemma B.2. *Let X_ϵ be a sequence in $L^\infty(\Omega, \mathscr{J}_1(\mathscr{H}))$. There is the weak* convergence $X_\epsilon \overset{*}{\rightharpoonup} X_0$ in $L^\infty(\Omega, \mathscr{J}_1(\mathscr{H}))$ if and only if both of the following conditions are satisfied:*

(i) *the sequence is bounded in $L^\infty(\Omega, \mathscr{J}_1(\mathscr{H}))$,*

(ii) *for all scalar functions $\chi \in L^1(\Omega)$ and elements $\phi, \psi \in \mathscr{H}$ there is the convergence*

$$\int_\Omega \chi(t)\langle X_\epsilon(t)\phi, \psi\rangle \, dt \;\rightarrow\; \int_\Omega \chi(t)\langle X_0(t)\phi, \psi\rangle \, dt.$$

Proof. By the very definition[68] of the Bochner integral, simple functions are dense in $L^1(\Omega, \mathscr{K}(\mathscr{H}))$. Since the operators of finite rank are dense in $\mathscr{K}(\mathscr{H})$, Eq. (B.2), we can restrict ourselves to simple functions having operators of finite rank as values. Thus, it suffices to test bounded sequences with all the functions of the form

$$Y = \chi \, \langle \cdot, \psi\rangle\phi \in L^1(\Omega, \mathscr{K}(\mathscr{H})),$$

where $\chi \in L^1(\Omega)$ is a characteristic function, and $\phi, \psi \in \mathscr{H}$. The necessity part of the proof follows as usual from the uniform boundedness principle, [83, Theorem 2.5]. □

Principle I.1 generalizes with virtually the same proof:

Lemma B.3. *Let $\{X_\epsilon\}$ be a sequence in $C^1(\overline{\Omega}, \mathscr{J}_1(\mathscr{H}))$ such that*

$$X_\epsilon \rightarrow 0 \quad \text{in} \quad C(\overline{\Omega}, \mathscr{J}_1(\mathscr{H})).$$

Then, if and only if the sequence $\{\partial X_\epsilon\}$ is bounded in $L^\infty(\Omega, \mathscr{J}_1(\mathscr{H}))$, there holds

$$\partial X_\epsilon \overset{*}{\rightharpoonup} 0 \quad \text{in} \quad L^\infty(\Omega, \mathscr{J}_1(\mathscr{H})).$$

Using the ideal properties, Eq. (B.4), we can also generalize Principle I.2 using virtually the same proof:

Lemma B.4. *Let there be the convergences $X_\epsilon \overset{*}{\rightharpoonup} X_0$ in $L^\infty(\Omega, \mathscr{J}_1(\mathscr{H}))$ and $Y_\epsilon \rightarrow Y_0$ in $L^\infty(\Omega, \mathscr{B}(\mathscr{H}))$. Then, there holds*

$$X_\epsilon \cdot Y_\epsilon \overset{*}{\rightharpoonup} X_0 \cdot Y_0, \quad Y_\epsilon \cdot X_\epsilon \overset{*}{\rightharpoonup} Y_0 \cdot X_0 \quad \text{in} \quad L^\infty(\Omega, \mathscr{J}_1(\mathscr{H})).$$

---

[68]Cf. [24, Definition II.2.1]

If we restrict the values of the test sequence Y_ϵ to be *compact* operators, we can prove the weak* convergence of the traces:

Lemma B.5. *Let there be the convergences* $X_\epsilon \overset{*}{\rightharpoonup} X_0$ *in* $L^\infty(\Omega, \mathcal{J}_1(\mathcal{H}))$ *and* $Y_\epsilon \to Y_0$ *in* $L^\infty(\Omega, \mathcal{K}(\mathcal{H}))$. *Then, there holds*

$$\mathrm{tr}(X_\epsilon \cdot Y_\epsilon) \overset{*}{\rightharpoonup} \mathrm{tr}(X_0 \cdot Y_0) \quad \text{in} \quad L^\infty(\Omega, \mathbb{C}).$$

Proof. Given a function $\chi \in L^1(\Omega, \mathbb{C})$, the Hölder inequality shows the strong convergence $\chi Y_\epsilon \to \chi Y_0$ in $L^1(\Omega, \mathcal{K}(\mathcal{H}))$. Theorem B.3 yields the convergence

$$\int_\Omega \chi(t) \cdot \mathrm{tr}(X_\epsilon(t) \cdot Y_\epsilon(t)) \, dt = \int_\Omega \mathrm{tr}(X_\epsilon(t) \cdot \chi(t) Y_\epsilon(t)) \, dt$$

$$\to \int_\Omega \mathrm{tr}(X_0(t) \cdot \chi(t) Y_0(t)) \, dt = \int_\Omega \chi(t) \cdot \mathrm{tr}(X_0(t) \cdot Y_0(t)) \, dt,$$

which proves the assertion. □

Further, one can show that weak*-convergence preserves certain structural information about the operator values.

Lemma B.6. *Assume that* $X_\epsilon \overset{*}{\rightharpoonup} X_0$ *in* $L^\infty(\Omega, \mathcal{J}_1(\mathcal{H}))$. *Then, there are the following implications:*

(i) *If the values of* X_ϵ *are nonnegative operators, a.e. in* Ω, *then the same holds for the limit* X_0.

(ii) *If the values of* X_ϵ *are selfadjoint operators, a.e. in* Ω, *then the same holds for the limit* X_0.

Proof. Suppose that the values of X_ϵ are *nonnegative* operators, a.e. in Ω. For a given element $\psi \in \mathcal{H}$ we define $Y \equiv \langle \cdot \psi \rangle \psi \in L^\infty(\Omega, \mathcal{K}(\mathcal{H}))$. We have

$$\chi_\epsilon(t) = \mathrm{tr}(X_\epsilon(t) \cdot Y(t)) = \langle X_\epsilon(t)\psi, \psi \rangle \geq 0, \quad \text{a.e. in } \Omega.$$

By Lemma B.5, the real-valued functions χ_ϵ converge weakly* in $L^\infty(\Omega, \mathbb{R})$ to χ_0,

$$\chi_0(t) = \mathrm{tr}(X_0(t) \cdot Y(t)) = \langle X_0(t)\psi, \psi \rangle.$$

Since Ω is bounded, this implies the *weak* convergence $\chi_\epsilon \rightharpoonup \chi_0$ in $L^2(\Omega, \mathbb{R})$. Here, the norm-closed convex cone of nonnegative functions is weakly closed. Thus, the limit χ_0 must be nonnegative. This proves the non-negativity of the values of X_0, a.e. in Ω.

Now, suppose that the values of X_ϵ are *selfadjoint* operators, a.e. in Ω. For given two elements $\phi, \psi \in \mathcal{H}$, we define $Y \equiv \langle \cdot, \psi \rangle \phi \in L^\infty(\Omega, \mathcal{K}(\mathcal{H}))$

and $Z \equiv \langle \cdot, \phi \rangle \psi \in L^\infty(\Omega, \mathscr{K}(\mathscr{H}))$. Since the values of X_ϵ are selfadjoint, a.e. in Ω, we obtain

$$\mathrm{tr}(X_\epsilon \cdot Y) = \langle X_\epsilon \phi, \psi \rangle = \overline{\langle X_\epsilon \psi, \phi \rangle} = \overline{\mathrm{tr}(X_\epsilon \cdot Z)},$$

Lemma B.5 allows to pass to the limit in $L^\infty(\Omega, \mathbb{C})$. Since complex conjugation is weakly*-continuous in $L^\infty(\Omega, \mathbb{C})$, we obtain

$$\langle X_0 \phi, \psi \rangle = \mathrm{tr}(X_0 \cdot Y) = \overline{\mathrm{tr}(X_0 \cdot Z)} = \overline{\langle X_0 \psi, \phi \rangle},$$

which implies that the values of X_0 are selfadjoint, a.e. in Ω.   □

§1.4.   The Trace is Not Weakly* Sequentially Continuous

According to the estimate (B.5), the trace extends naturally as a linear operator

$$\mathrm{tr} : L^\infty(\Omega, \mathscr{J}_1(\mathscr{H})) \to L^\infty(\Omega, \mathbb{C}),$$

continuous in the norm topologies. This implies continuity in the weak topology [25, Theorem V.3.15], but not, however, in the weak* topology. In fact, if $\mathscr{H}$ is infinite dimensional, the trace is *not* weakly* sequentially continuous—which is the source of considerable difficulties in Chapter IV.[69]

Lemma B.7. Let $\mathscr{H}$ be an infinite dimensional Hilbert space with an orthonormal basis $\{\phi_n\}$. The sequence $X_n \equiv \langle \cdot, \phi_n \rangle \phi_n$ fulfills

$$X_n \xrightarrow{*} 0 \quad \text{in} \quad L^\infty(\Omega, \mathscr{J}_1(\mathscr{H})), \qquad \text{but} \qquad \mathrm{tr}\, X_n \equiv 1 \quad \text{in} \quad L^\infty(\Omega, \mathbb{C}).$$

Proof. Since the trace class norm of $X_n(t)$ is uniformly bounded,

$$\|X_n(\cdot)\|_{\mathscr{J}_1} = \mathrm{tr}\, X_n(\cdot) = 1,$$

we get by Theorem B.3 that there is a subsequence n' and a limit $X_\infty \in L^\infty(\Omega, \mathscr{J}_1(\mathscr{H}))$, such that

$$X_{n'} \xrightarrow{*} X_\infty \quad \text{in} \quad L^\infty(\Omega, \mathscr{J}_1(\mathscr{H})).$$

For a given vector $\psi \in \mathscr{H}$, we test this weak*-convergence with the function

$$Y = \langle \cdot, X_\infty \psi \rangle \psi \in L^1(\Omega, \mathscr{K}(\mathscr{H})).$$

and obtain that

$$\int_\Omega \langle X_{n'}(t)\psi, X_\infty(t)\psi \rangle \, dt = \int_\Omega \mathrm{tr}(X_{n'}(t) \cdot Y(t)) \, dt$$

$$\to \int_\Omega \mathrm{tr}(X_\infty(t) \cdot Y(t)) \, dt = \int_\Omega \|X_\infty(t)\psi\|^2 \, dt.$$

---

[69]Lemma B.5 would yield the weak*-continuity of the trace, *if* the identity operator on $\mathscr{H}$ were compact. However, the identity operator is compact if and only if $\mathscr{H}$ is finite dimensional.

Parseval's equality shows that an orthonormal basis converges weakly to zero, $\phi_n \rightharpoonup 0$. Thus, we get

$$\langle X_n(t)\psi, X_\infty(t)\psi \rangle = \langle \psi, \phi_n \rangle \langle \phi_n, X_\infty(t)\psi \rangle \ \rightarrow \ 0, \qquad \text{a.e. in } \Omega.$$

Hence, by Lebesgue's theorem of dominated convergence, we obtain

$$\int_\Omega \|X_\infty(t)\psi\|^2 \, dt = 0,$$

which implies $X_\infty = 0$. By Principle I.5, we can discard the extraction of subsequences and have thus proven that $X_n \xrightarrow{*} 0$ in $L^\infty(\Omega, \mathscr{J}_1(\mathscr{H}))$. $\qquad \square$

§1.5.   Trace Class Operators on Rigged Hilbert Spaces

In Chapter IV we have to work in a scale of Hilbert spaces rather than a single space. Here, we will discuss the underlying relation of the corresponding trace class operators.

Let there be a separable Hilbert $\mathscr{H}_+$, *densely* embedded in $\mathscr{H}$,

$$\mathscr{H}_+ \hookrightarrow \mathscr{H}.$$

The corresponding inner product will be denoted by $\langle \cdot, \cdot \rangle_{\mathscr{H}_+}$. Each element $\phi \in \mathscr{H}$ defines by $\psi \mapsto \langle \psi, \phi \rangle$ a continuous linear form on $\mathscr{H}_+$. That way one obtains a dense embedding,[70]

$$\mathscr{H} \hookrightarrow \mathscr{H}_- = \mathscr{H}_+^*,$$

into the dual space $\mathscr{H}_-$ of $\mathscr{H}_+$, being a separable Hilbert space as well. Using density and uniform continuity arguments, the inner product $\langle \cdot, \cdot \rangle$ of $\mathscr{H}$ extends uniquely to an sesquilinear form on $\mathscr{H}_+ \times \mathscr{H}_-$, giving the dual pairing of the two spaces which is *conjugate* linear in the second argument. The norm

$$\|\phi\|_{\mathscr{H}_-} = \sup_{\psi \in \mathscr{H}_+} \frac{\langle \phi, \psi \rangle}{\|\psi\|_{\mathscr{H}_+}}, \qquad \phi \in \mathscr{H}_-,$$

fulfills the parallelogram equality and stems thus from an inner product $\langle \cdot, \cdot \rangle_{\mathscr{H}_-}$ on $\mathscr{H}_-$. Any such triple

$$\mathscr{H}_+ \hookrightarrow \mathscr{H} \hookrightarrow \mathscr{H}_-$$

of Hilbert spaces is called a *rigging* of $\mathscr{H}$, or sometimes, a *Gelfand triple*. Having such a triple in mind, the space $\mathscr{H}$ is called a *rigged Hilbert space*. The Riesz representation theorem, [78, Theorem II.4], shows that there is an *unitary* operator

$$\sigma : \mathscr{H}_- \to \mathscr{H}_+,$$

---

[70]The Hahn-Banach separation theorem, [83, Theorem 3.4], shows that $\mathscr{H} \hookrightarrow \mathscr{H}_-$ is an embedding because $\mathscr{H}_+ \subset \mathscr{H}$ is dense, and it is dense because $\mathscr{H}_+ \hookrightarrow \mathscr{H}$ is an embedding.

defined by

$$\langle \phi, \sigma\psi \rangle_{\mathscr{H}_+} = \langle \phi, \psi \rangle, \qquad \phi \in \mathscr{H}_+, \ \psi \in \mathscr{H}_-.$$

The *inverse* of σ is given by the adjoint operator

$$\sigma^* : \mathscr{H}_+ \to \mathscr{H}_-,$$

which satisfies

$$\langle \sigma^*\phi, \psi \rangle_{\mathscr{H}_-} = \langle \phi, \psi \rangle, \qquad \phi \in \mathscr{H}_+, \ \psi \in \mathscr{H}_-.$$

This map σ will be called the *Riesz representation map* of the rigging. The natural embedding $\mathscr{H}_+ \hookrightarrow \mathscr{H}_-$ will be abbreviated by ι.

The Riesz map σ provides a convenient way to view operators defined on $\mathscr{H}_-$ as operators on $\mathscr{H}_+$ and vice versa: For $A \in \mathscr{B}(\mathscr{H}_-)$ we define

$$A^\sigma = A\iota\sigma \in \mathscr{B}(\mathscr{H}_-), \qquad {}^\sigma A = \sigma A\iota \in \mathscr{B}(\mathscr{H}_+).$$

Likewise, for $A \in \mathscr{B}(\mathscr{H}_+)$ we define

$$A^\sigma = \iota A\sigma \in \mathscr{B}(\mathscr{H}_-), \qquad {}^\sigma A = \sigma\iota A \in \mathscr{B}(\mathscr{H}_+).$$

These mappings restrict to trace class operators.

Lemma B.8. *The mapping $A \mapsto A^\sigma$ is norm continuous from $\mathscr{J}_1(\mathscr{H}_-)$ to $\mathscr{J}_1(\mathscr{H}_-)$, and from $\mathscr{J}_1(\mathscr{H}_+)$ to $\mathscr{J}_1(\mathscr{H}_-)$. Correspondingly, the mapping $A \mapsto {}^\sigma A$ is norm continuous from $\mathscr{J}_1(\mathscr{H}_-)$ to $\mathscr{J}_1(\mathscr{H}_+)$, and from $\mathscr{J}_1(\mathscr{H}_+)$ to $\mathscr{J}_1(\mathscr{H}_+)$.*

Proof. Let $A \in \mathscr{J}_1(\mathscr{H}_-)$ be given. By the ideal property, Eq. (B.4), we obtain

$$\|A^\sigma\|_{\mathscr{J}_1(\mathscr{H}_-)} \le \|A\|_{\mathscr{J}_1(\mathscr{H}_-)} \|\iota\sigma\|_{\mathscr{B}(\mathscr{H}_-)} \le \|\iota\| \cdot \|A\|_{\mathscr{J}_1(\mathscr{H}_-)}.$$

Taking the singular value decomposition

$$A = \sum_k \lambda_k \langle \cdot, \psi_k \rangle_{\mathscr{H}_-} \phi_k$$

we obtain

$${}^\sigma A = \left(\sum_k \lambda_k \langle \cdot, \sigma\psi_k \rangle_{\mathscr{H}_+} \sigma\phi_k \right) \circ \sigma\iota.$$

Since $\{\sigma\phi_k\}$ and $\{\sigma\psi_k\}$ are orthonormal in $\mathscr{H}_+$, the first term of the right hand side represents the singular value decomposition of an operator in $\mathscr{J}_1(\mathscr{H}_+)$ yielding

$$\|{}^\sigma A\|_{\mathscr{J}_1(\mathscr{H}_+)} \le \sum_k \lambda_k \cdot \|\sigma\iota\|_{\mathscr{B}(\mathscr{H}_+)} \le \|\iota\| \cdot \|A\|_{\mathscr{J}_1(\mathscr{H}_-)}.$$

Now, let $A \in \mathscr{J}_1(\mathscr{H}_+)$ be given. Again, the ideal property, Eq. (B.4), yields

$$\|^\sigma A\|_{\mathscr{J}_1(\mathscr{H}_+)} \leq \|\sigma\iota\|_{\mathscr{B}(\mathscr{H}_+)} \|A\|_{\mathscr{J}_1(\mathscr{H}_+)} \leq \|\iota\| \cdot \|A\|_{\mathscr{J}_1(\mathscr{H}_+)}.$$

The singular value decomposition

$$A = \sum_k \lambda_k \langle \cdot, \psi_k \rangle_{\mathscr{H}_+} \phi_k$$

transforms as

$$A^\sigma = \iota\sigma \circ \left(\sum_k \lambda_k \langle \cdot, \sigma^*\psi_k \rangle_{\mathscr{H}_-} \sigma^*\phi_k \right).$$

Since $\{\sigma^*\phi_k\}$ and $\{\sigma^*\psi_k\}$ are orthonormal in $\mathscr{H}_-$, the second term of the right hand side represents the singular value decomposition of an operator in $\mathscr{J}_1(\mathscr{H}_-)$ yielding

$$\|A^\sigma\|_{\mathscr{J}_1(\mathscr{H}_+)} \leq \|\iota\sigma\|_{\mathscr{B}(\mathscr{H}_-)} \sum_k \lambda_k \leq \|\iota\| \cdot \|A\|_{\mathscr{J}_1(\mathscr{H}_+)},$$

which completes the proof.   □

The thus introduced maps extend naturally to norm continuous maps on the spaces $L^\infty(\Omega, \mathscr{J}_1(\mathscr{H}_\pm))$. This does not automatically imply that these maps are weakly* sequentially continuous, as the counterexample of §B.1.4 has shown. However, nothing pathological happens here.

Lemma B.9. *Assume that $X_\epsilon \overset{*}{\rightharpoonup} X_0$ in $L^\infty(\Omega, \mathscr{J}_1(\mathscr{H}_\pm))$. Then, there is*

$$X_\epsilon^\sigma \overset{*}{\rightharpoonup} X_0^\sigma \quad \text{in} \quad L^\infty(\Omega, \mathscr{J}_1(\mathscr{H}_-)), \quad {}^\sigma X_\epsilon \overset{*}{\rightharpoonup} {}^\sigma X_0 \quad \text{in} \quad L^\infty(\Omega, \mathscr{J}_1(\mathscr{H}_+)).$$

Proof. Consider a sequence $X_\epsilon \overset{*}{\rightharpoonup} X_0$ in $L^\infty(\Omega, \mathscr{J}_1(\mathscr{H}_-))$. Lemma B.4 yields

$$X_\epsilon^\sigma = X_\epsilon \iota\sigma \overset{*}{\rightharpoonup} X_0 \iota\sigma = X_0^\sigma \quad \text{in} \quad L^\infty(\Omega, \mathscr{J}_1(\mathscr{H}_-)).$$

Because of the continuity in norm, the sequence ${}^\sigma X_\epsilon$ is bounded in the space $L^\infty(\Omega, \mathscr{J}_1(\mathscr{H}_+))$. For $\chi \in L^1(\Omega)$ and $\phi, \psi \in \mathscr{H}_+$ we obtain

$$\begin{aligned}
\int_\Omega \chi(t) \langle {}^\sigma X_\epsilon(t)\phi, \psi \rangle_{\mathscr{H}_+} dt &= \int_\Omega \chi(t) \langle \sigma X_\epsilon(t)\phi, \psi \rangle_{\mathscr{H}_+} dt \\
&= \int_\Omega \chi(t) \langle X_\epsilon(t)\phi, \sigma^*\psi \rangle_{\mathscr{H}_-} dt \\
&\rightarrow \int_\Omega \chi(t) \langle X_0(t)\phi, \sigma^*\psi \rangle_{\mathscr{H}_-} dt \\
&= \int_\Omega \chi(t) \langle {}^\sigma X_0(t)\phi, \psi \rangle_{\mathscr{H}_+} dt.
\end{aligned}$$

Hence, Lemma B.2 implies $^\sigma X_\epsilon \overset{*}{\to} {}^\sigma X_0$ in $L^\infty(\Omega, \mathscr{J}_1(\mathscr{H}_+))$.

Now, consider a sequence $X_\epsilon \overset{*}{\to} X_0$ in $L^\infty(\Omega, \mathscr{J}_1(\mathscr{H}_+))$. By Lemma B.4 we get

$$^\sigma X_\epsilon = \sigma \iota\, X_\epsilon \overset{*}{\to} \sigma \iota\, X_0 = {}^\sigma X_0 \quad \text{in} \quad L^\infty(\Omega, \mathscr{J}_1(\mathscr{H}_+)).$$

Because of the continuity in norm, the sequence X_ϵ^σ is bounded in the space $L^\infty(\Omega, \mathscr{J}_1(\mathscr{H}_-))$. For $\chi \in L^1(\Omega)$ and $\phi, \psi \in \mathscr{H}_-$ we obtain

$$
\begin{aligned}
\int_\Omega \chi(t)\langle X_\epsilon^\sigma(t)\phi, \psi\rangle_{\mathscr{H}_-}\, dt
&= \int_\Omega \chi(t)\langle X_\epsilon(t)\sigma\phi, \psi\rangle_{\mathscr{H}_-}\, dt \\
&= \int_\Omega \chi(t)\langle{}^\sigma X_\epsilon(t)\sigma\phi, \sigma\psi\rangle_{\mathscr{H}_+}\, dt \\
&\to \int_\Omega \chi(t)\langle{}^\sigma X_0(t)\sigma\phi, \sigma\psi\rangle_{\mathscr{H}_+}\, dt \\
&= \int_\Omega \chi(t)\langle X_0^\sigma(t)\phi, \psi\rangle_{\mathscr{H}_-}\, dt.
\end{aligned}
$$

Hence, Lemma B.2 implies $X_\epsilon^\sigma \overset{*}{\to} X_0^\sigma$ in $L^\infty(\Omega, \mathscr{J}_1(\mathscr{H}_-))$.   □

We finish our study of trace-class-operator valued functions by the following lemma.

Lemma B.10. *Let there be a sequence* $\{X_\epsilon\}$ *in* $L^\infty(\Omega, \mathscr{B}(\mathscr{H}_-, \mathscr{H}_+))$ *with the following three properties:*

(i) $X_\epsilon \sigma^* \overset{*}{\to} X_0^+$ *in* $L^\infty(\Omega, \mathscr{J}_1(\mathscr{H}_+))$,

(ii) $\{X_\epsilon|_{\mathscr{H}}\}$ *is bounded in* $L^\infty(\Omega, \mathscr{J}_1(\mathscr{H}))$,

(iii) $\{\sigma^* X_\epsilon\}$ *is bounded in* $L^\infty(\Omega, \mathscr{J}_1(\mathscr{H}_-))$.

Then, there is $X_\epsilon|_{\mathscr{H}} \overset{*}{\to} X_0|_{\mathscr{H}}$ *in* $L^\infty(\Omega, \mathscr{J}_1(\mathscr{H}))$ *and* $\sigma^* X_\epsilon \overset{*}{\to} X_0^-$ *in* $L^\infty(\Omega, \mathscr{J}_1(\mathscr{H}_-))$ *with*

$$X_0 = X_0^+ \sigma = \sigma X_0^- \in L^\infty(\Omega, \mathscr{B}(\mathscr{H}_-, \mathscr{H}_+)).$$

Proof. Let $\chi \in L^1(\Omega)$ be given. For $\phi, \psi \in \mathscr{H}$ we obtain

$$
\begin{aligned}
\int_\Omega \chi(t)\langle X_\epsilon(t)\phi, \psi\rangle\, dt
&= \int_\Omega \chi(t)\langle X_\epsilon(t)\sigma^* \cdot \sigma\phi, \sigma\psi\rangle_{\mathscr{H}_+}\, dt \\
&\to \int_\Omega \chi(t)\langle X_0^+(t) \cdot \sigma\phi, \sigma\psi\rangle_{\mathscr{H}_+}\, dt \\
&= \int_\Omega \chi(t)\langle X_0^+(t)\sigma\phi, \psi\rangle\, dt.
\end{aligned}
$$

By Lemma B.2, this shows $X_0 = X_0^+\sigma$. On the other hand, for $\phi, \psi \in \mathscr{H}_-$ we obtain

$$
\begin{aligned}
\int_\Omega \chi(t)\langle \sigma^* X_\epsilon(t)\phi, \psi\rangle_{\mathscr{H}_-}\, dt &= \int_\Omega \chi(t)\langle X_\epsilon(t)\sigma^* \cdot \sigma\phi, \sigma\psi\rangle_{\mathscr{H}_+}\, dt \\
&\to \int_\Omega \chi(t)\langle X_0^+(t) \cdot \sigma\phi, \sigma\psi\rangle_{\mathscr{H}_+}\, dt \\
&= \int_\Omega \chi(t)\langle \sigma^* X_0^+(t)\sigma\phi, \psi\rangle_{\mathscr{H}_-}\, dt.
\end{aligned}
$$

By Lemma B.2, this shows $X_0^- = \sigma^* X_0^+ \sigma$.   □

§2.   Semi-Bounded Operators and Coercive Quadratic Forms

Quadratic forms present a convenient way for constructing Hamiltonians in quantum mechanics; an excellent account of this technique is provided by the textbook [91] of SIMON. We restrict ourselves to the very basic facts which in a way are variations on the two themes "Friedrichs extension" and "Lax-Milgram theorem."

Given a rigged Hilbert space $\mathscr{H}$ with the rigging $\mathscr{H}_+ \hookrightarrow \mathscr{H} \hookrightarrow \mathscr{H}_-$, as defined in §B.1.5, we consider quadratic forms

$$
a : \mathscr{H}_+ \times \mathscr{H}_+ \to \mathbb{C},
$$

linear in the first, conjugate linear in the second argument. The space $\mathscr{H}_+$ is called the *form domain* of a. The quadratic form a is called *symmetric*, if

$$
a(\phi, \psi) = \overline{a(\psi, \phi)}, \qquad \phi, \psi \in \mathscr{H}_+.
$$

The quadratic form a is called $\mathscr{H}_+$-coercive, if it is continuous and there are constants $\gamma > 0$ and $\kappa \in \mathbb{R}$ such that

$$
a(\phi, \phi) \geq \gamma\langle \phi, \phi\rangle_{\mathscr{H}_+} - \kappa\langle \phi, \phi\rangle.
$$

Notice, that a coercive form is automatically symmetric if $\mathscr{H}$ is complex. This way, the inner product $a(\phi, \psi) + \kappa\langle \phi, \psi\rangle$ becomes equivalent to the original inner product $\langle \phi, \psi\rangle_{\mathscr{H}_+}$ of $\mathscr{H}_+$.

There is a one-to-one correspondence of semi-bounded selfadjoint operators and coercive quadratic forms on $\mathscr{H}$ which we will describe now: Given a coercive form a, there is a *unique* selfadjoint operator $A : D(A) \subset \mathscr{H} \to \mathscr{H}$ with domain $D(A) \subset \mathscr{H}_+$ having the following two properties:

(i) A extends to a *bounded* linear operator $A : \mathscr{H}_+ \to \mathscr{H}_-$,

(ii) A represents the form a, i.e, $a(\phi, \psi) = \langle A\phi, \psi\rangle$ for all $\phi, \psi \in \mathscr{H}_+$.

The selfadjoint operator A is semi-bounded,

$$\langle A\phi, \phi \rangle \geq -\kappa \|\phi\|^2, \qquad \phi \in D(A).$$

We say that A is *associated* to a and call $\mathcal{H}_+$ the *form domain of A*. On the other hand, given a semi-bounded selfadjoint operator A on $\mathcal{H}$, there exists a unique rigging $\mathcal{H}_+ \hookrightarrow \mathcal{H} \hookrightarrow \mathcal{H}_-$, and a unique H_+-coercive quadratic form a associated to A. Proofs for all these claims can be found in [76, §VIII.6].

In a way which we do not explain here, one can associate a specific symmetric quadratic form a to a given selfadjoint operator A, such that

$$a(\phi, \psi) = \langle A\phi, \psi \rangle, \qquad \phi, \psi \in D(A).$$

The form domain of a will be called the form domain of A as well. However, without semi-boundedness, A does not extend to the form domain $\mathcal{H}_+$ in general. Nevertheless, one *formally* writes $\langle A\phi, \psi \rangle = a(\phi, \psi)$ if $\phi, \psi \in \mathcal{H}_+$. Details can be found in [76, Example VIII.6.2].

An important tool for constructing selfadjoint operators is given by the "KLMN theorem," the letters of the acronym standing for KATO, LIONS, LAX, MILGRAM, and NELSON, cf. [76, Theorem X.17].

Theorem B.4. *Let A be a nonnegative selfadjoint operator on $\mathcal{H}$ with form domain $\mathcal{H}_+$. Suppose that B is a selfadjoint operator with a form domain containing $\mathcal{H}_+$ such that*

$$|\langle B\phi, \phi \rangle| \leq \alpha \langle A\phi, \phi \rangle + \beta \langle \phi, \phi \rangle, \qquad \phi \in \mathcal{H}_+,$$

for some $\alpha < 1$ and $\beta \in \mathbb{R}$. Then there exists a unique selfadjoint operator C with form domain $\mathcal{H}_+$ and

$$\langle C\phi, \psi \rangle = \langle A\phi, \psi \rangle + \langle B\phi, \psi \rangle \qquad \phi, \psi \in \mathcal{H}_+.$$

The operator C is semi-bounded with bound $-\beta$, i.e.,

$$\langle C\phi, \phi \rangle \geq -\beta \|\phi\|^2, \qquad \phi \in D(C).$$

One writes for short: $C = A + B$, defined as the sum of quadratic forms.

§2.1. Rollnik-Potentials

A measurable function $V : \mathbb{R}^3 \to \mathbb{C}$, obeying

$$\|V\|_{\mathscr{R}}^2 = \int_{\mathbb{R}^3} \int_{\mathbb{R}^3} \frac{|V(x)|\,|V(y)|}{|x - y|^2}\, dx\, dy \, < \, \infty,$$

belongs to the *Rollnik class $\mathscr{R}$*. This class is a Banach-space with norm $\|\cdot\|_{\mathscr{R}}$, cf. [91, §I.2]. There is the continuous embedding

$$L^{3/2}(\mathbb{R}^3) \hookrightarrow \mathscr{R},$$

cf. [91, Theorem I.1]. For instance, the central potential $V(x) = -|x|^{-\alpha}$ fulfills $V \in \mathscr{R} + L^\infty(\mathbb{R}^3)$ if and only if $0 < \alpha < 2$ which is the physically reasonable bound. This fact is one of the main reasons for considering potentials in $\mathscr{R} + L^\infty(\mathbb{R}^3)$ instead of just $L^2(\mathbb{R}^3) + L^\infty(\mathbb{R}^3)$.

Now, if $V \in \mathscr{R} + L^\infty(\mathbb{R}^3)$, one can define the two-body Hamiltonian

$$H = -\Delta + V,$$

as the sum of quadratic forms by referring to the KLMN theorem, cf. [91, Corollary II.8]. The thus obtained Hamiltonian is semi-bounded, selfadjoint on $L^2(\mathbb{R}^3)$ with form domain $H^1(\mathbb{R}^3)$. The corresponding rigging of $L^2(\mathbb{R}^3)$ is thus given by the Sobolev spaces,

$$H^1(\mathbb{R}^3) \hookrightarrow L^2(\mathbb{R}^3) \hookrightarrow H^{-1}(\mathbb{R}^3),$$

and the unbounded selfadjoint operator H extends as a *bounded* linear operator to

$$H : H^1(\mathbb{R}^3) \to H^{-1}(\mathbb{R}^3).$$

For the purposes of Chapter IV, we have to study a parameter-dependent situation. Let $\Omega \subset \mathbb{R}^d$ be a bounded domain. Given functions

$$V_1 \in C^k(\overline{\Omega}, \mathscr{R}), \qquad V_2 \in C^k(\overline{\Omega}, L^\infty(\mathbb{R}^3)),$$

we define the parameter dependent Hamiltonian

$$H(t) = -\Delta + V_1(t) + V_2(t), \qquad t \in \Omega,$$

as above as the sum of quadratic forms. Going through the details of the proof of [91, Corollary II.8], one obtains the following facts: The family of selfadjoint operators $H(t)$ extends to the Sobolev space $H^1(\mathbb{R}^3)$ in such a way that

$$H \in C^k\left(\overline{\Omega}, \mathscr{B}\left(H^1(\mathbb{R}^3), H^{-1}(\mathbb{R}^3)\right)\right).$$

The associated quadratic form $h(t; \phi, \psi)$ is *uniformly H^1-coercive*, i.e., there are constants $\gamma > 0$ and $\kappa \in \mathbb{R}$ such that

$$h(t; \phi, \phi) \geq \gamma \|\phi\|_{H^1}^2 - \kappa \|\phi\|_{L^2}^2 \qquad \forall t \in \Omega, \, \forall \phi \in H^1(\mathbb{R}^3).$$

Appendix C:
Asymptotic Studies of Two Model Problems

In this part of the appendix we study the singular limit of two model problems *asymptotically*. We will use the tools of the *perturbation theory of integrable Hamiltonian systems* as they are presented, for instance, in the survey of ARNOLD, KOZLOV, and NEISHTADT [6, Chap. 5]. This asymptotic study here serves a twofold reason. First, we want to show the fundamental differences in method and scope between an asymptotic approach and that approach we have pursued in this monograph: being oblivious to the microscales by using weak convergences. Second, we want to convince the reader that our method is much easier to handle, especially in the presence of resonances. Even for these rather simple model problems formulas easily mess up and, to the best of the author's knowledge, no-one has ever tried to follow them in the more general case which has been subject of Chapter II.

§1. The Model Problem of the Introduction

To begin, we will discuss the introductory model problem of §I.2.1 using the notation introduced there. Additionally, we will denote by (η, ζ) the canonical momenta corresponding to the positions (y, z). This way, the equations of motion, Eq. (I.4), together with the velocity relations

$$\dot{y}_\epsilon = \eta_\epsilon, \qquad \dot{z}_\epsilon = \zeta_\epsilon,$$

are given by the canonical equations of motion belonging to the energy function

$$E_\epsilon = \tfrac{1}{2}|\eta_\epsilon|^2 + \tfrac{1}{2}|\zeta_\epsilon|^2 + \tfrac{1}{2}\epsilon^{-2}\sum_\lambda \omega_\lambda^2(y_\epsilon)(z_\epsilon^\lambda)^2.$$

We introduce particular action-angle variables $(\theta_\epsilon, \phi_\epsilon)$ for the fast, normal degrees of freedom $(z_\epsilon, \zeta_\epsilon)$,[71]

$$z_\epsilon^\lambda = \epsilon\sqrt{\frac{2\theta_\epsilon^\lambda}{\omega_\lambda(y_\epsilon)}}\,\sin(\epsilon^{-1}\phi_\epsilon^\lambda), \qquad \zeta_\epsilon^\lambda = \sqrt{2\theta_\epsilon^\lambda\omega_\lambda(y_\epsilon)}\,\cos(\epsilon^{-1}\phi_\epsilon^\lambda).$$

---

[71]Such a particular choice presumes quite a substantial amount of insight into the underlying problem.

This transformation yields the one-forms

$$
dz_\epsilon^\lambda = \sqrt{\frac{2\theta_\epsilon^\lambda}{\omega_\lambda(y_\epsilon)}} \cos(\epsilon^{-1}\phi_\epsilon^\lambda)\, d\phi_\epsilon^\lambda \; + \; \epsilon\sqrt{\frac{1}{2\theta_\epsilon^\lambda\omega_\lambda(y_\epsilon)}} \sin(\epsilon^{-1}\phi_\epsilon^\lambda)\, d\theta_\epsilon^\lambda
$$

$$
- \; \epsilon\sum_j \partial_j\omega_\lambda(y_\epsilon)\sqrt{\frac{\theta_\epsilon^\lambda}{2\omega_\lambda^3(y_\epsilon)}} \sin(\epsilon^{-1}\phi_\epsilon^\lambda)\, dy_\epsilon^j,
$$

and

$$
d\zeta_\epsilon^\lambda = - \; \epsilon^{-1}\sqrt{2\theta_\epsilon^\lambda\omega_\lambda(y_\epsilon)}\, \sin(\epsilon^{-1}\phi_\epsilon^\lambda)\, d\phi_\epsilon^\lambda \; + \; \sqrt{\frac{\omega_\lambda(y_\epsilon)}{2\theta_\epsilon^\lambda}} \cos(\epsilon^{-1}\phi_\epsilon^\lambda)\, d\theta_\epsilon^\lambda
$$

$$
+ \; \sum_j \partial_j\omega_\lambda(y_\epsilon)\sqrt{\frac{\theta_\epsilon^\lambda}{2\omega_\lambda(y_\epsilon)}} \cos(\epsilon^{-1}\phi_\epsilon^\lambda)\, dy_\epsilon^j.
$$

Hence, we obtain

$$
\sum_\lambda dz_\epsilon^\lambda \wedge d\zeta_\epsilon^\lambda = \sum_\lambda d\phi_\epsilon^\lambda \wedge d\theta_\epsilon^\lambda \; + \; \sum_{\lambda,j} \theta_\epsilon^\lambda \frac{\partial_j\omega_\lambda(y_\epsilon)}{\omega_\lambda(y_\epsilon)} \cos(2\epsilon^{-1}\phi_\epsilon^\lambda)\, d\phi_\epsilon^\lambda \wedge dy_\epsilon^j
$$

$$
+ \; \epsilon\sum_{\lambda,j} \frac{\partial_j\omega_\lambda(y_\epsilon)}{2\omega_\lambda(y_\epsilon)} \sin(2\epsilon^{-1}\phi_\epsilon^\lambda)\, d\theta_\epsilon^\lambda \wedge dy_\epsilon^j
$$

$$
- \; \epsilon\sum_{\lambda,j,k} \theta_\epsilon^\lambda \frac{\partial_j\omega_\lambda(y_\epsilon)\cdot\partial_k\omega_\lambda(y_\epsilon)}{2\omega_\lambda^2(y_\epsilon)} \sin(\epsilon^{-1}\phi_\epsilon^\lambda)\cos(\epsilon^{-1}\phi_\epsilon^\lambda)\, dy_\epsilon^j \wedge dy_\epsilon^k.
$$

Symmetry considerations show that the last term is *zero*. However, for obtaining a transformation being symplectic on the phase-space as a whole, we additionally have to transform the remaining momenta,

$$
\eta_\epsilon^j = p_\epsilon^j + \epsilon\sum_\lambda \frac{\theta_\epsilon^\lambda\cdot\partial_j\omega_\lambda(y_\epsilon)}{2\omega_\lambda(y_\epsilon)} \sin(2\epsilon^{-1}\phi_\epsilon^\lambda).
$$

This transformation results in the one-form

$$
d\eta_\epsilon^j = dp_\epsilon^j + \sum_\lambda \frac{\theta_\epsilon^\lambda\cdot\partial_j\omega_\lambda(y_\epsilon)}{\omega_\lambda(y_\epsilon)} \cos(2\epsilon^{-1}\phi_\epsilon^\lambda)\, d\phi_\epsilon^\lambda
$$

$$
+ \; \epsilon\sum_\lambda \frac{\partial_j\omega_\lambda(y_\epsilon)}{2\omega_\lambda(y_\epsilon)} \sin(2\epsilon^{-1}\phi_\epsilon^\lambda)\, d\theta_\epsilon^\lambda
$$

$$
+ \; \epsilon\sum_{\lambda,k} \theta_\epsilon^\lambda \left(\frac{\partial_k\partial_j\omega_\lambda(y_\epsilon)}{2\omega_\lambda(y_\epsilon)} - \frac{\partial_j\omega_\lambda(y_\epsilon)\cdot\partial_k\omega_\lambda(y_\epsilon)}{2\omega_\lambda^2(y_\epsilon)} \right) \sin(2\epsilon^{-1}\phi_\epsilon^\lambda)\, dy_\epsilon^k,
$$

and therefore

$$\sum_j dy_\epsilon^j \wedge d\eta_\epsilon^j = \sum_j dy_\epsilon^j \wedge dp_\epsilon^j - \sum_{\lambda,j} \theta_\epsilon^\lambda \frac{\partial_j \omega_\lambda(y_\epsilon)}{\omega_\lambda(y_\epsilon)} \cos(2\epsilon^{-1}\phi_\epsilon^\lambda) \, d\phi_\epsilon^\lambda \wedge dy_\epsilon^j$$

$$- \epsilon \sum_{\lambda,j} \frac{\partial_j \omega_\lambda(y_\epsilon)}{2\omega_\lambda(y_\epsilon)} \sin(2\epsilon^{-1}\phi_\epsilon^\lambda) \, d\theta_\epsilon^\lambda \wedge dy_\epsilon^j$$

$$+ \epsilon \sum_{\lambda,j,k} \theta_\epsilon^\lambda \left(\frac{\partial_k \partial_j \omega_\lambda(y_\epsilon)}{2\omega_\lambda(y_\epsilon)} - \frac{\partial_j \omega_\lambda(y_\epsilon) \cdot \partial_k \omega_\lambda(y_\epsilon)}{2\omega_\lambda^2(y_\epsilon)} \right) \sin(2\epsilon^{-1}\phi_\epsilon^\lambda) \, dy_\epsilon^j \wedge dy_\epsilon^k.$$

Again, for reasons of symmetry, the last term is *zero*. Altogether, these lengthy but straightforward calculations have proven that the transformation $(y_\epsilon, \eta_\epsilon; z_\epsilon, \zeta_\epsilon) \mapsto (y_\epsilon, p_\epsilon; \phi_\epsilon, \theta_\epsilon)$ is symplectic indeed,

$$\sum_j dy_\epsilon^j \wedge d\eta_\epsilon^j + \sum_\lambda dz_\epsilon^\lambda \wedge d\zeta_\epsilon^\lambda = \sum_j dy_\epsilon^j \wedge dp_\epsilon^j + \sum_\lambda d\phi_\epsilon^\lambda \wedge d\theta_\epsilon^\lambda.$$

The energy transforms to the expression

$$E_\epsilon = \tfrac{1}{2}|p_\epsilon|^2 + \sum_\lambda \theta_\epsilon^\lambda \omega_\lambda(y_\epsilon) + \epsilon \sum_{j,\lambda} \frac{p_\epsilon^j \theta_\epsilon^\lambda \cdot \partial_j \omega_\lambda(y_\epsilon)}{2\omega_\lambda(y_\epsilon)} \sin(2\epsilon^{-1}\phi_\epsilon^\lambda)$$

$$+ \tfrac{1}{8}\epsilon^2 \sum_j \left(\sum_\lambda \frac{\theta_\epsilon^\lambda \cdot \partial_j \omega_\lambda(y_\epsilon)}{\omega_\lambda(y_\epsilon)} \sin(2\epsilon^{-1}\phi_\epsilon^\lambda) \right)^2.$$

Thus, by the canonical formalism, the equation of motion take the form

$$\dot\phi_\epsilon^\lambda = \frac{\partial E_\epsilon}{\partial \theta_\epsilon^\lambda}, \qquad \dot\theta_\epsilon^\lambda = -\frac{\partial E_\epsilon}{\partial \phi_\epsilon^\lambda}, \qquad \dot y_\epsilon^j = \frac{\partial E_\epsilon}{\partial p_\epsilon^j}, \qquad \dot p_\epsilon^j = -\frac{\partial E_\epsilon}{\partial y_\epsilon^j},$$

i.e., after some calculation,

$$\dot\phi_\epsilon^\lambda = \omega_\lambda(y_\epsilon) + \epsilon \sum_j \frac{p_\epsilon^j \cdot \partial_j \omega_\lambda(y_\epsilon)}{2\omega_\lambda(y_\epsilon)} \sin(2\epsilon^{-1}\phi_\epsilon^\lambda) + O(\epsilon^2),$$

$$\dot\theta_\epsilon^\lambda = -\sum_j \frac{p_\epsilon^j \theta_\epsilon^\lambda \cdot \partial_j \omega_\lambda(y_\epsilon)}{\omega_\lambda(y_\epsilon)} \cos(2\epsilon^{-1}\phi_\epsilon^\lambda) + \epsilon \sum_\mu F_{\lambda\mu}^1(y_\epsilon, p_\epsilon, \theta_\epsilon)$$
$$\cdot \left(\sin\left(2\epsilon^{-1}(\phi_\epsilon^\mu - \phi_\epsilon^\lambda)\right) + \sin\left(2\epsilon^{-1}(\phi_\epsilon^\mu + \phi_\epsilon^\lambda)\right) \right),$$

$$\dot y_\epsilon^j = p_\epsilon^j + \epsilon \sum_\lambda \frac{\theta_\epsilon^\lambda \cdot \partial_j \omega_\lambda(y_\epsilon)}{2\omega_\lambda(y_\epsilon)} \sin(2\epsilon^{-1}\phi_\epsilon^\lambda),$$

$$\dot p_\epsilon^j = -\sum_\lambda \theta_\epsilon^\lambda \cdot \partial_j \omega_\lambda(y_\epsilon) + \epsilon \sum_\lambda F_{j\lambda}^2(y_\epsilon, p_\epsilon, \theta_\epsilon) \sin(2\epsilon^{-1}\phi_\epsilon^\lambda) + O(\epsilon^2).$$

Here, we abbreviate by $F^1_{\lambda\mu}(y,p,\theta)$ and $F^2_{j\lambda}(y,p,\theta)$ certain smooth expressions in the variables y,p and θ. The initial values, as given in Eq. (I.5), transform to

$$\phi_\epsilon(0) = 0, \qquad \theta^\lambda_\epsilon(0) = \theta^\lambda_* = \frac{|u^\lambda_*|^2}{2\omega_\lambda(y_*)}, \qquad y_\epsilon(0) = y_*, \qquad p_\epsilon(0) = w_*.$$

Perturbation theory now proceeds by eliminating the fast dependence on the angle variables of the equations of motion for increasing order of ϵ. We will see that we can eliminate the dependence of the $O(1)$-terms in general, but that a resonance condition will be necessary for eliminating the dependence of the $O(\epsilon)$-terms.

A First Order Approximation. For eliminating the fast dependence on the angle variables of the $O(1)$-terms we introduce the transformed action variables

$$\Theta^\lambda_\epsilon = \theta^\lambda_\epsilon + \epsilon \sum_j \frac{p^j_\epsilon \theta^\lambda_\epsilon \cdot \partial_j \omega_\lambda(y_\epsilon)}{2\omega^2_\lambda(y_\epsilon)} \sin(2\epsilon^{-1}\phi^\lambda_\epsilon), \qquad \Theta_\epsilon(0) = \theta_*.$$

Now, the equations of motion take the simple form

$$\dot\phi^\lambda_\epsilon = \omega_\lambda(y_\epsilon) + O(\epsilon),$$
$$\dot\Theta^\lambda_\epsilon = O(\epsilon),$$
$$\dot y^j_\epsilon = p^j_\epsilon + O(\epsilon),$$
$$\dot p^j_\epsilon = -\sum_\lambda \Theta^\lambda_\epsilon \cdot \partial_j \omega_\lambda(y_\epsilon) + O(\epsilon).$$

A comparison with the limit system

$$\begin{aligned}
\dot\phi^\lambda_0 &= \omega_\lambda(y_0), & \phi_0(0) &= 0, \\
\dot\Theta^\lambda_0 &= 0, & \Theta_0(0) &= \theta_*, \\
\dot y^j_0 &= p^j_0, & y_0(0) &= y_*, \\
\dot p^j_0 &= -\sum_\lambda \Theta^\lambda_0 \cdot \partial_j \omega_\lambda(y_0), & p_0(0) &= w_*,
\end{aligned} \qquad \text{(C.1)}$$

reveals, for times $t = O(1)$, that there are the asymptotic estimates

$$\phi_\epsilon = \phi_0 + O(\epsilon), \quad \Theta_\epsilon = \theta_* + O(\epsilon), \quad y_\epsilon = y_0 + O(\epsilon), \quad p_\epsilon = \dot y_0 + O(\epsilon),$$

i.e., after transforming back,

$$\theta_\epsilon = \theta_* + O(\epsilon), \qquad y_\epsilon = y_0 + O(\epsilon), \qquad \dot y_\epsilon = \dot y_0 + O(\epsilon).$$

We observe that the equations (C.1) are just the *homogenized* equations of motion as introduced in Theorem I.1, i.e., there holds $y_0 = y_{\text{hom}}$. This way, we have given a second, independent proof of Theorem I.1 with an additional error estimate. However, the reader should notice that the argument is not as straightforward as in §I.2 and the calculations are much more involved.

A Second Order Approximation. Additional elimination of the dependence on the angle variables in the $O(\epsilon)$-terms requires the assumption that there are *no* resonances of order two,

$$\omega_\lambda(y) \neq \omega_\mu(y), \qquad y \in \mathbb{R}^n, \ \lambda \neq \mu.$$

This assumptions enables us to transform the action variables to

$$\Theta_\epsilon^\lambda = \theta_\epsilon^\lambda + \epsilon \sum_j \frac{p_\epsilon^j \theta_\epsilon^\lambda \cdot \partial_j \omega_\lambda(y_\epsilon)}{2\omega_\lambda^2(y_\epsilon)} \sin(2\epsilon^{-1}\phi_\epsilon^\lambda) + \epsilon^2 \sum_{\mu \neq \lambda} F_{\lambda\mu}^1(y_\epsilon, p_\epsilon, \theta_\epsilon)$$

$$\cdot \frac{1}{2} \left(\frac{\cos\left(2\epsilon^{-1}(\phi_\epsilon^\mu - \phi_\epsilon^\lambda)\right)}{\omega_\mu(y_\epsilon) - \omega_\lambda(y_\epsilon)} + \frac{\cos\left(2\epsilon^{-1}(\phi_\epsilon^\mu + \phi_\epsilon^\lambda)\right)}{\omega_\mu(y_\epsilon) + \omega_\lambda(y_\epsilon)} \right)$$

with initial value

$$\Theta_\epsilon(0) = \theta_* + O(\epsilon^2).$$

Likewise, we transform the angle variables to

$$\Phi_\epsilon^\lambda = \phi_\epsilon^\lambda + \epsilon^2 \sum_j \frac{p_\epsilon^j \cdot \partial_j \omega_\lambda(y_\epsilon)}{4\omega_\lambda^2(y_\epsilon)} \cos(2\epsilon^{-1}\phi_\epsilon^\lambda), \qquad \Phi_\epsilon(0) = O(\epsilon^2).$$

This way we get the following equations of motion:

$$\dot{\Phi}_\epsilon^\lambda = \omega_\lambda(y_\epsilon) + O(\epsilon^2),$$

$$\dot{\Theta}_\epsilon^\lambda = \epsilon F_\lambda^3(y_\epsilon, p_\epsilon, \Theta_\epsilon) \sin(2\epsilon^{-1}\Phi_\epsilon^\lambda) + \epsilon F_\lambda^4(y_\epsilon, p_\epsilon, \Theta_\epsilon) \sin(4\epsilon^{-1}\Phi_\epsilon^\lambda) + O(\epsilon^2),$$

$$\dot{y}_\epsilon^j = p_\epsilon^j + \epsilon \sum_\lambda F_{j\lambda}^5(y_\epsilon, p_\epsilon, \Theta_\epsilon) \sin(2\epsilon^{-1}\Phi_\epsilon^\lambda) + O(\epsilon^2),$$

$$\dot{p}_\epsilon^j = -\sum_\lambda \Theta_\epsilon^\lambda \cdot \omega_\lambda(y_\epsilon) + \epsilon \sum_\lambda F_{j\lambda}^6(y_\epsilon, p_\epsilon, \Theta_\epsilon) \sin(2\epsilon^{-1}\Phi_\epsilon^\lambda) + O(\epsilon^2).$$

Here, we use $F_\lambda^3(y, p, \Theta)$, $F_\lambda^4(y, p, \Theta)$, $F_{j\lambda}^5(y, p, \Theta)$, and $F_{j\lambda}^6(y, p, \Theta)$ to abbreviate certain expressions smoothly dependent on y, p, and Θ. This system is now in such a form that we can completely eliminate the fast dependence on the angle variables of the $O(\epsilon)$-terms. Upon introducing the transformed variables

$$I_\epsilon^\lambda = \Theta_\epsilon^\lambda + \epsilon^2 \frac{F_\lambda^3(y_\epsilon, p_\epsilon, \Theta_\epsilon)}{2\omega_\lambda(y_\epsilon)} \cos(2\epsilon^{-1}\Phi_\epsilon^\lambda) + \epsilon^2 \frac{F_\lambda^4(y_\epsilon, p_\epsilon, \Theta_\epsilon)}{4\omega_\lambda(y_\epsilon)} \cos(4\epsilon^{-1}\Phi_\epsilon^\lambda),$$

$$Y_\epsilon^j = y_\epsilon^j + \epsilon^2 \sum_\lambda \frac{F_{j\lambda}^5(y_\epsilon, p_\epsilon, \Theta_\epsilon)}{2\omega_\lambda(y_\epsilon)} \cos(2\epsilon^{-1}\Phi_\epsilon^\lambda),$$

$$P_\epsilon^j = p_\epsilon^j + \epsilon^2 \sum_\lambda \frac{F_{j\lambda}^6(y_\epsilon, p_\epsilon, \Theta_\epsilon)}{2\omega_\lambda(y_\epsilon)} \cos(2\epsilon^{-1}\Phi_\epsilon^\lambda),$$

with initial values

$$I_\epsilon(0) = \theta_* + O(\epsilon^2), \qquad Y_\epsilon(0) = y_* + O(\epsilon^2), \qquad P_\epsilon(0) = w_* + O(\epsilon^2),$$

we obtain the following simple equations of motion:

$$\dot{\Phi}_\epsilon^\lambda = \omega_\lambda(Y_\epsilon) + O(\epsilon^2),$$

$$\dot{I}_\epsilon^\lambda = O(\epsilon^2),$$

$$\dot{Y}_\epsilon^j = P_\epsilon^j + O(\epsilon^2),$$

$$\dot{P}_\epsilon^j = -\sum_\lambda I_\epsilon^\lambda \cdot \partial_j \omega_\lambda(Y_\epsilon) + O(\epsilon^2).$$

A comparison with the limit system of Eq. (C.1) yields, for times $t = O(1)$, the estimates

$$\Phi_\epsilon = \phi_0 + O(\epsilon^2), \quad I_\epsilon = \theta_* + O(\epsilon^2), \quad Y_\epsilon = y_0 + O(\epsilon^2), \quad P_\epsilon = \dot{y}_0 + O(\epsilon^2).$$

Expressing these estimates in the original variables we have finally proven that

$$\phi_\epsilon = \phi_0 + O(\epsilon^2),$$

$$\theta_\epsilon^\lambda = \theta_*^\lambda \left(1 - \epsilon \sum_j \frac{\dot{y}_0^j \cdot \partial_j \omega_\lambda(y_0)}{2\omega_\lambda^2(y_0)} \sin(2\epsilon^{-1}\phi_0^\lambda) \right) + O(\epsilon^2),$$

$$y_\epsilon = y_0 + O(\epsilon^2),$$

$$\dot{y}_\epsilon^j = \dot{y}_0^j + \epsilon \sum_\lambda \frac{\theta_*^\lambda \cdot \partial_j \omega_\lambda(y_0)}{2\omega_\lambda(y_0)} \sin(2\epsilon^{-1}\phi_0^\lambda) + O(\epsilon^2),$$

$$z_\epsilon^\lambda = \epsilon \sqrt{\frac{2\theta_*^\lambda}{\omega_\lambda(y_0)}} \sin(\epsilon^{-1}\phi_0^\lambda) + O(\epsilon^2),$$

$$\dot{z}_\epsilon^\lambda = \sqrt{2\theta_*^\lambda \omega_\lambda(y_0)} \cos(\epsilon^{-1}\phi_0^\lambda) + O(\epsilon).$$

Remark C.1. These asymptotic expressions nicely show that there is a strong convergence $\dot{z}_\epsilon \to 0$, or $\epsilon^{-1} z_\epsilon \to 0$, if and only if $\theta_* = 0$. The latter condition is equivalent to $u_* = 0$, i.e., that the initial velocity is *tangential* to the constraint manifold $N = \{x = (y, z) : z = 0\}$. This observation is in perfect accordance with Lemma II.17. In the case $\theta_* = 0$ the estimates above improve to

$$\theta_\epsilon = \theta_* + O(\epsilon^2), \qquad y_\epsilon = y_0 + O(\epsilon^2), \qquad \dot{y}_\epsilon = \dot{y}_0 + O(\epsilon^2).$$

These latter estimates can also be found in the work of LUBICH [66].

§2.   Quantum-Classical Coupling (Finite Dimensional)

To end with, we will discuss the finite dimensional version of the quantum-classical coupling model of §III.3 using the notation introduced there. This time, we define the phase-space coordinates $(y_\epsilon, \eta_\epsilon; z_\epsilon, \zeta_\epsilon)$ by[72]

$$\psi_\epsilon = \frac{\epsilon^{-1} z_\epsilon + i\zeta_\epsilon}{\sqrt{2}}, \qquad \eta_\epsilon = \dot{y}_\epsilon.$$

The quantum-classical coupling equations are the canonical equations of motion, Eq. (III.9), belonging to the energy function

$$E_\epsilon = \tfrac{1}{2}|\eta_\epsilon|^2 + \tfrac{1}{2}\langle H(y_\epsilon)\zeta_\epsilon, \zeta_\epsilon\rangle + \tfrac{1}{2}\epsilon^{-2}\langle H(y_\epsilon)z_\epsilon, z_\epsilon\rangle.$$

We will assume right from the beginning that all eigenvalues of $H(y)$ are simple and that there are no resonances of order two,

$$\omega_\lambda(y) \neq \omega_\mu(y), \qquad y \in \mathbb{R}^n, \ \lambda \neq \mu.$$

Because $H(y)$ is assumed to be *real* symmetric, there is a family of *real* orthonormal eigenvectors $(e_1(y), \ldots, e_r(y))$,

$$H(y)e_\lambda(y) = \omega_\lambda(y)e_\lambda(y), \qquad \langle e_\lambda(y), e_\mu(y)\rangle = \delta_{\lambda\mu}.$$

This normalization yields important anti-symmetry relations of the derivatives $\partial_j e_\lambda(y)$, specifically

$$\langle e_\lambda(y), \partial_j e_\mu(y)\rangle = -\langle e_\mu(y), \partial_j e_\lambda(y)\rangle,$$

$$\langle \partial_j e_\lambda(y), \partial_k e_\mu(y)\rangle = -\langle \partial_k e_\lambda(y), \partial_j e_\mu(y)\rangle \qquad \text{(C.2)}$$

$$+ \text{ terms symmetric in } j \text{ and } k.$$

As before in §C.1.1, we introduce particular action-angle variables $(\theta_\epsilon, \phi_\epsilon)$ for the fast, normal degrees of freedom $(z_\epsilon, \zeta_\epsilon)$,

$$z_\epsilon = \epsilon \sum_\lambda \sqrt{2\theta_\epsilon^\lambda} \cos(\epsilon^{-1}\phi_\epsilon^\lambda)\, e_\lambda(y_\epsilon), \quad \zeta_\epsilon = -\sum_\lambda \sqrt{2\theta_\epsilon^\lambda} \sin(\epsilon^{-1}\phi_\epsilon^\lambda)\, e_\lambda(y_\epsilon).$$

This transformation yields the one-forms

$$\begin{aligned}
dz_\epsilon^\lambda = & -\sum_\mu \sqrt{2\theta_\epsilon^\mu} \sin(\epsilon^{-1}\phi_\epsilon^\mu)\, e_\mu^\lambda(y_\epsilon)\, d\phi_\epsilon^\mu \\
& + \epsilon \sum_\mu \frac{1}{\sqrt{2\theta_\epsilon^\mu}} \cos(\epsilon^{-1}\phi_\epsilon^\mu)\, e_\mu^\lambda(y_\epsilon)\, d\theta_\epsilon^\mu \\
& + \epsilon \sum_{\mu,j} \sqrt{2\theta_\epsilon^\mu} \cos(\epsilon^{-1}\phi_\epsilon^\mu)\, \partial_j e_\mu^\lambda(y_\epsilon)\, dy_\epsilon^j,
\end{aligned}$$

and

$$\begin{aligned}
d\zeta_\epsilon^\lambda = & -\epsilon^{-1} \sum_\mu \sqrt{2\theta_\epsilon^\mu} \cos(\epsilon^{-1}\phi_\epsilon^\mu)\, e_\mu^\lambda(y_\epsilon)\, d\phi_\epsilon^\mu \\
& - \sum_\mu \frac{1}{\sqrt{2\theta_\epsilon^\mu}} \sin(\epsilon^{-1}\phi_\epsilon^\mu)\, e_\mu^\lambda(y_\epsilon)\, d\theta_\epsilon^\mu \\
& - \sum_{\mu,j} \sqrt{2\theta_\epsilon^\mu} \sin(\epsilon^{-1}\phi_\epsilon^\mu)\, \partial_j e_\mu^\lambda(y_\epsilon)\, dy_\epsilon^j.
\end{aligned}$$

---

[72]I.e., we ignore the constant γ of Eq. (III.8) which has been introduced for getting particular initial values in §III.3 only.

Hence, by using the normalization $\langle e_\lambda, e_\mu \rangle = \delta_{\lambda\mu}$, we obtain

$$
\begin{aligned}
\sum_\lambda dz_\epsilon^\lambda \wedge d\zeta_\epsilon^\lambda = &\sum_\lambda d\phi_\epsilon^\lambda \wedge d\theta_\epsilon^\lambda \\
&+ \sum_{\lambda,\mu,j} 2\sqrt{\theta_\epsilon^\lambda \theta_\epsilon^\mu} \cos\left(\epsilon^{-1}(\phi_\epsilon^\lambda - \phi_\epsilon^\mu)\right) \langle e_\lambda(y_\epsilon), \partial_j e_\mu(y_\epsilon)\rangle \, d\phi_\epsilon^\lambda \wedge dy_\epsilon^j \\
&+ \epsilon \sum_{\lambda,\mu,j} \sqrt{\frac{\theta_\epsilon^\mu}{\theta_\epsilon^\lambda}} \sin\left(\epsilon^{-1}(\phi_\epsilon^\lambda - \phi_\epsilon^\mu)\right) \langle e_\lambda(y_\epsilon), \partial_j e_\mu(y_\epsilon)\rangle \, d\theta_\epsilon^\lambda \wedge dy_\epsilon^j \\
&+ \epsilon \sum_{\lambda,\mu,j,k} \sqrt{\theta_\epsilon^\lambda \theta_\epsilon^\mu} \sin\left(\epsilon^{-1}(\phi_\epsilon^\lambda - \phi_\epsilon^\mu)\right) \langle \partial_j e_\lambda(y_\epsilon), \partial_k e_\mu(y_\epsilon)\rangle \, dy_\epsilon^j \wedge dy_\epsilon^k.
\end{aligned}
$$

However, for obtaining a transformation being symplectic on the phase-space as a whole, we additionally have to transform the remaining momenta,

$$
\eta_\epsilon^j = p_\epsilon^j + \epsilon \sum_{\lambda,\mu} \sqrt{\theta_\epsilon^\lambda \theta_\epsilon^\mu} \sin\left(\epsilon^{-1}(\phi_\epsilon^\lambda - \phi_\epsilon^\mu)\right) \langle e_\lambda(y_\epsilon), \partial_j e_\mu(y_\epsilon)\rangle. \tag{C.3}
$$

By the first of the anti-symmetry relations in Eq. (C.2), this transformation results in the one-form

$$
\begin{aligned}
d\eta_\epsilon^j = d p_\epsilon^j \; + &\sum_{\lambda,\mu} 2\sqrt{\theta_\epsilon^\lambda \theta_\epsilon^\mu} \cos\left(\epsilon^{-1}(\phi_\epsilon^\lambda - \phi_\epsilon^\mu)\right) \langle e_\lambda(y_\epsilon), \partial_j e_\mu(y_\epsilon)\rangle \, d\phi_\epsilon^\lambda \\
&+ \epsilon \sum_{\lambda,\mu} \sqrt{\frac{\theta_\epsilon^\mu}{\theta_\epsilon^\lambda}} \sin\left(\epsilon^{-1}(\phi_\epsilon^\lambda - \phi_\epsilon^\mu)\right) \langle e_\lambda(y_\epsilon), \partial_j e_\mu(y_\epsilon)\rangle \, d\theta_\epsilon^\lambda \\
&+ \epsilon \sum_{\lambda,\mu,k} \sqrt{\theta_\epsilon^\lambda \theta_\epsilon^\mu} \sin\left(\epsilon^{-1}(\phi_\epsilon^\lambda - \phi_\epsilon^\mu)\right) \big(\langle \partial_k e_\lambda(y_\epsilon), \partial_j e_\mu(y_\epsilon)\rangle \\
&\qquad\qquad + \langle e_\lambda(y_\epsilon), \partial_k \partial_j e_\mu(y_\epsilon)\rangle\big) \, dy_\epsilon^k.
\end{aligned}
$$

Hence, by using the second of the anti-symmetry relations in Eq. (C.2), we have

$$
\begin{aligned}
\sum_j dy_\epsilon^j \wedge d\eta_\epsilon^j = &\sum_j dy_\epsilon^j \wedge d p_\epsilon^j \\
&- \sum_{\lambda,\mu,j} 2\sqrt{\theta_\epsilon^\lambda \theta_\epsilon^\mu} \cos\left(\epsilon^{-1}(\phi_\epsilon^\lambda - \phi_\epsilon^\mu)\right) \langle e_\lambda(y_\epsilon), \partial_j e_\mu(y_\epsilon)\rangle \, d\phi_\epsilon^\lambda \wedge dy_\epsilon^j \\
&- \epsilon \sum_{\lambda,\mu,j} \sqrt{\frac{\theta_\epsilon^\mu}{\theta_\epsilon^\lambda}} \sin\left(\epsilon^{-1}(\phi_\epsilon^\lambda - \phi_\epsilon^\mu)\right) \langle e_\lambda(y_\epsilon), \partial_j e_\mu(y_\epsilon)\rangle \, d\theta_\epsilon^\lambda \wedge dy_\epsilon^j \\
&- \epsilon \sum_{\lambda,\mu,j,k} \sqrt{\theta_\epsilon^\lambda \theta_\epsilon^\mu} \sin\left(\epsilon^{-1}(\phi_\epsilon^\lambda - \phi_\epsilon^\mu)\right) \langle \partial_j e_\lambda(y_\epsilon), \partial_k e_\mu(y_\epsilon)\rangle \, dy_\epsilon^j \wedge dy_\epsilon^k.
\end{aligned}
$$

Altogether, these lengthy but straightforward calculations have proven that the transformation $(y_\epsilon, \eta_\epsilon; z_\epsilon, \zeta_\epsilon) \mapsto (y_\epsilon, p_\epsilon; \phi_\epsilon, \theta_\epsilon)$ is symplectic indeed,

$$\sum_j dy_\epsilon^j \wedge d\eta_\epsilon^j + \sum_\lambda dz_\epsilon^\lambda \wedge d\zeta_\epsilon^\lambda = \sum_j dy_\epsilon^j \wedge dp_\epsilon^j + \sum_\lambda d\phi_\epsilon^\lambda \wedge d\theta_\epsilon^\lambda.$$

The energy transforms to the expression

$$E_\epsilon = \tfrac{1}{2}|p_\epsilon|^2 + \sum_\lambda \theta_\epsilon^\lambda \omega_\lambda(y_\epsilon)$$

$$+ \epsilon \sum_{\lambda,\mu,j} p_\epsilon^j \sqrt{\theta_\epsilon^\lambda \theta_\epsilon^\mu} \sin\left(\epsilon^{-1}(\phi_\epsilon^\lambda - \phi_\epsilon^\mu)\right) \langle e_\lambda(y_\epsilon), \partial_j e_\mu(y_\epsilon)\rangle$$

$$+ \tfrac{1}{2}\epsilon^2 \sum_j \left(\sum_{\lambda,\mu} \sqrt{\theta_\epsilon^\lambda \theta_\epsilon^\mu} \sin\left(\epsilon^{-1}(\phi_\epsilon^\lambda - \phi_\epsilon^\mu)\right) \langle e_\lambda(y_\epsilon), \partial_j e_\mu(y_\epsilon)\rangle\right)^2.$$

Thus, by the canonical formalism, the equation of motion take the form

$$\dot\phi_\epsilon^\lambda = \frac{\partial E_\epsilon}{\partial \theta_\epsilon^\lambda}, \qquad \dot\theta_\epsilon^\lambda = -\frac{\partial E_\epsilon}{\partial \phi_\epsilon^\lambda}, \qquad \dot y_\epsilon^j = \frac{\partial E_\epsilon}{\partial p_\epsilon^j}, \qquad \dot p_\epsilon^j = -\frac{\partial E_\epsilon}{\partial y_\epsilon^j},$$

i.e., after some calculation,

$$\dot\phi_\epsilon^\lambda = \omega_\lambda(y_\epsilon) + \epsilon \sum_{\mu \neq \lambda, j} p_\epsilon^j \sqrt{\frac{\theta_\epsilon^\mu}{\theta_\epsilon^\lambda}} \sin\left(\epsilon^{-1}(\phi_\epsilon^\lambda - \phi_\epsilon^\mu)\right) \langle e_\lambda(y_\epsilon), \partial_j e_\mu(y_\epsilon)\rangle$$

$$+ O(\epsilon^2),$$

$$\dot\theta_\epsilon^\lambda = -2 \sum_{\mu \neq \lambda, j} p_\epsilon^j \sqrt{\theta_\epsilon^\lambda \theta_\epsilon^\mu} \cos\left(\epsilon^{-1}(\phi_\epsilon^\lambda - \phi_\epsilon^\mu)\right) \langle e_\lambda(y_\epsilon), \partial_j e_\mu(y_\epsilon)\rangle$$

$$+ \epsilon \sum_{\substack{\mu_1, \mu_2 \\ \mu_1 \neq \mu_2}} F^1_{\lambda\mu_1\mu_2}(y_\epsilon, p_\epsilon, \theta_\epsilon) \sin\left(\epsilon^{-1}(\phi_\epsilon^{\mu_1} - \phi_\epsilon^{\mu_2})\right)$$

$$+ \epsilon \sum_{\substack{\mu_1, \mu_2, \mu_3 \\ \lambda \neq \mu_2, \lambda \neq \mu_3 \\ \mu_1 \neq \mu_2, \mu_1 \neq \mu_3}} F^2_{\lambda\mu_1\mu_2\mu_3}(y_\epsilon, p_\epsilon, \theta_\epsilon) \sin\left(\epsilon^{-1}(\phi_\epsilon^\lambda + \phi_\epsilon^{\mu_1} - \phi_\epsilon^{\mu_2} - \phi_\epsilon^{\mu_3})\right),$$

$$\dot y_\epsilon^j = p_\epsilon^j + \epsilon \sum_{\lambda,\mu} \sqrt{\theta_\epsilon^\lambda \theta_\epsilon^\mu} \sin\left(\epsilon^{-1}(\phi_\epsilon^\lambda - \phi_\epsilon^\mu)\right) \langle e_\lambda(y_\epsilon), \partial_j e_\mu(y_\epsilon)\rangle,$$

$$\dot p_\epsilon^j = -\sum_\lambda \theta_\epsilon^\lambda \cdot \partial_j \omega_\lambda(y_\epsilon) + \epsilon \sum_{\lambda,\mu} F^3_{j\lambda\mu}(y_\epsilon, p_\epsilon, \theta_\epsilon) \sin\left(\epsilon^{-1}(\phi_\epsilon^\lambda - \phi_\epsilon^\mu)\right)$$

$$+ O(\epsilon^2).$$

Here, we abbreviate by $F^1_{\lambda\mu_1\mu_2}(y, p, \theta)$, $F^2_{\lambda\mu_1\mu_2\mu_3}(y, p, \theta)$, and $F^3_{j\lambda\mu}(y, p, \theta)$ certain smooth expressions in the variables y, p and θ. The initial values, as given in Eq. (III.6), transform as follows. First, we represent the initial excitations using polar coordinates,

$$\langle \psi_*, e_\lambda(y_*) \rangle = \sqrt{\theta_*^\lambda} \cdot \exp\left(-i\phi_*^\lambda\right), \qquad \lambda = 1, \ldots, r.$$

Next, we obtain

$$\phi_\epsilon(0) = \epsilon\phi_*, \qquad \theta_\epsilon(0) = \theta_*, \qquad y_\epsilon(0) = y_*, \qquad p_\epsilon(0) = w_* + O(\epsilon).$$

A First Order Approximation. Now, for eliminating the fast dependence on the angle variables of the $O(1)$-terms we introduce the transformed action variables

$$\Theta_\epsilon^\lambda = \theta_\epsilon^\lambda + 2\epsilon \sum_{\mu \neq \lambda, j} \frac{p_\epsilon^j \sqrt{\theta_\epsilon^\lambda \theta_\epsilon^\mu}}{\omega_\lambda(y_\epsilon) - \omega_\mu(y_\epsilon)} \sin\left(\epsilon^{-1}(\phi_\epsilon^\lambda - \phi_\epsilon^\mu)\right) \langle e_\lambda(y_\epsilon), \partial_j e_\mu(y_\epsilon) \rangle,$$

$$\text{(C.4)}$$

with initial value $\Theta_\epsilon(0) = \theta_* + O(\epsilon)$. Since we have excluded any resonance of order two, this transformation is well-defined. The equations of motion take the simple form

$$\dot{\phi}_\epsilon^\lambda = \omega_\lambda(y_\epsilon) + O(\epsilon),$$
$$\dot{\Theta}_\epsilon^\lambda = O(\epsilon),$$
$$\dot{y}_\epsilon^j = p_\epsilon^j + O(\epsilon),$$
$$\dot{p}_\epsilon^j = -\sum_\lambda \Theta_\epsilon^\lambda \cdot \partial_j \omega_\lambda(y_\epsilon) + O(\epsilon).$$

A comparison with the limit system

$$\dot{\phi}_0^\lambda = \omega_\lambda(y_0), \qquad\qquad \phi_0(0) = 0,$$
$$\dot{\Theta}_0^\lambda = 0, \qquad\qquad \Theta_0(0) = \theta_*,$$
$$\dot{y}_0^j = p_0^j, \qquad\qquad y_0(0) = y_*, \qquad\qquad \text{(C.5)}$$
$$\dot{p}_0^j = -\sum_\lambda \theta_*^\lambda \cdot \partial_j \omega_\lambda(y_0), \qquad p_0(0) = w_*,$$

reveals, for times $t = O(1)$, that there are the asymptotic estimates

$$\phi_\epsilon = \phi_0 + O(\epsilon), \quad \Theta_\epsilon = \theta_* + O(\epsilon), \quad y_\epsilon = y_0 + O(\epsilon), \quad p_\epsilon = \dot{y}_0 + O(\epsilon),$$

i.e., after transforming back,

$$\theta_\epsilon = \theta_* + O(\epsilon), \qquad y_\epsilon = y_0 + O(\epsilon), \qquad \dot{y}_\epsilon = \dot{y}_0 + O(\epsilon).$$

We observe that Eq. (C.5) is just another form of the *time-dependent Born-Oppenheimer model* which has been introduced in Theorem III.1. Thus, there holds $y_0 = y_{\text{hom}}$.

A Second Order Approximation. Elimination of the term

$$\epsilon \sum F^2_{\lambda\mu_1\mu_2\mu_3}(y_\epsilon, p_\epsilon, \theta_\epsilon) \sin\left(\epsilon^{-1}(\phi^\lambda_\epsilon + \phi^{\mu_1}_\epsilon - \phi^{\mu_2}_\epsilon - \phi^{\mu_3}_\epsilon)\right),$$

which appears in the differential equation for the action variables, requires the assumption that there are *no symmetric resonances of order four*, i.e.,

$$\omega_{\mu_1}(y) + \omega_{\mu_2}(y) \neq \omega_{\mu_3}(y) + \omega_{\mu_4}(y), \qquad y \in \mathbb{R}^n,$$

for

$$\mu_1 \neq \mu_3, \quad \mu_1 \neq \mu_4, \quad \mu_2 \neq \mu_3, \quad \mu_2 \neq \mu_4.$$

In fact, this assumptions enables us to eliminate *all* the $O(\epsilon)$-terms. We omit the details, which are completely analogously to what we have done in §C.1.1. However, recalling Eqs. (C.3) and (C.4), it should be clear that we finally obtain the asymptotic estimates

$$\phi_\epsilon = \phi_0 + O(\epsilon^2),$$

$$\theta^\lambda_\epsilon = \Theta^\lambda_* - 2\epsilon \sum_{\mu \neq \lambda, j} \frac{\dot{y}^j_0 \sqrt{\theta^\lambda_* \theta^\mu_*}}{\omega_\lambda(y_0) - \omega_\mu(y_0)} \sin\left(\epsilon^{-1}(\phi^\lambda_0 - \phi^\mu_0)\right) \langle e_\lambda(y_0), \partial_j e_\mu(y_0)\rangle$$
$$+ O(\epsilon^2),$$

$$y_\epsilon = y_0 + O(\epsilon^2),$$

$$\dot{y}^j_\epsilon = \dot{y}^j_0 + \epsilon \sum_{\lambda, \mu} \sqrt{\theta^\lambda_* \theta^\mu_*} \sin\left(\epsilon^{-1}(\phi^\lambda_0 - \phi^\mu_0)\right) \langle e_\lambda(y_0), \partial_j e_\mu(y_0)\rangle + O(\epsilon^2),$$

$$\psi_\epsilon = \sum_\lambda \sqrt{\theta^\lambda_*} \exp\left(-i\epsilon^{-1}\phi^\lambda_0\right) e_\lambda(y_0) + O(\epsilon).$$

This time, however, we have to deal with the $O(\epsilon)$-perturbations of the initial values explicitly. To be specific, the constants Θ_* are given by the values

$$\Theta^\lambda_* = \theta^\lambda_* + 2\epsilon \sum_{\mu \neq \lambda, j} \frac{w^j_* \sqrt{\theta^\lambda_* \theta^\mu_*}}{\omega_\lambda(y_*) - \omega_\mu(y_*)} \sin\left(\phi^\lambda_* - \phi^\mu_*\right) \langle e_\lambda(y_*), \partial_j e_\mu(y_*)\rangle.$$

Further, y_0 and ϕ_0 are defined as the solutions of the time-dependent Born-Oppenheimer model,

$$\ddot{y}_0 = -\sum_\lambda \Theta^\lambda_* \cdot \operatorname{grad} \omega_\lambda(y_0), \qquad \dot{\phi}^\lambda_0 = \omega_\lambda(y_0),$$

with the initial values $\phi_0(0) = \epsilon\phi_*$, $y_0(0) = y_*$, and

$$\dot{y}^j_0(0) = w^j_* - \epsilon \sum_{\lambda, \mu} \sqrt{\theta^\lambda_* \theta^\mu_*} \sin\left(\phi^\lambda_* - \phi^\mu_*\right) \langle e_\lambda(y_*), \partial_j e_\mu(y_*)\rangle.$$

Remark C.2. There are three immediate observations. First, if the initial wave function ψ_* is *real*, i.e., the initial vector of angles satisfies $\phi_* = 0$, the constants and initial values above simplify to

$$\Theta_* = \theta_*, \qquad \phi_0(0) = 0, \qquad \dot{y}_0(0) = w_*.$$

Second, if just *one* eigenstate, for instance the ground-state, is excited initially, we obtain the yet improved estimates

$$\theta_\epsilon = \Theta_* + O(\epsilon^2), \qquad \dot{y}_\epsilon = \dot{y}_0 + O(\epsilon^2).$$

Finally, in the course of this asymptotic study only *differences* of the frequencies ω_λ have appeared in the denominators of the expressions. Thus, there is no need to assume that H is uniformly positive definite. This is in perfect accordance with the more general results of Chapter IV.

Appendix D:
The Weak Virial Theorem and Localization
of Semiclassical Measures

In this final part of the appendix we sketch the relation of our results to the advanced analytical tools developed for the study of oscillations and concentration effects in nonlinear partial differential equations, in particular to cope with nonlinear expressions of only weakly converging sequences. This way we give our results a broader perspective and, hopefully, clarify the possible impact of these tools.

Let $x_\epsilon \rightharpoonup x_0$ be a sequence of vector-valued functions that depend on an argument $t \in \Omega \subset \mathbb{R}^d$. The weak convergence is understood to hold in $L^2(\Omega, \mathbb{R}^m)$. For describing the weak limits of nonlinear expression in x_ϵ, two major tools have been invented:

- *Young measure* [28, §1.E.3] describes, or better, encodes *all* limits of nonlinear substitutions $f(x_\epsilon)$, $f \in C(\mathbb{R}^m)$. The difficulty with this tool lies in the fact, that besides existence there is not much known about how to systematically obtain information on this measure.

- *H-measure* invented by TARTAR [95][96] and, independently, by GÉRARD [37] who uses the name *microlocal defect measure* instead. This measure encodes all limits of certain *quadratic pseudodifferential* expression of x_ϵ. There is a systematic way, called *localization principle*, how to obtain knowledge about this measure from any differential equation involving x_ϵ. For differential equations that explicitly involve a scale-parameter ϵ a variant of those measures applies, called *semiclassical measure* [36][97] or *Wigner measure* [65].

Thus, the tool of choice for the study of the quadratic quantities in Chapter II is the semiclassical measure. We will show that the localization principle obtained from the differential equation (II.25) just yields the weak virial theorem, Eq. (II.27).

To begin, we recall the basic properties of the semiclassical measure as introduced by GÉRARD [36, Prop. 3.1 and Eqs. (3.11), (3.12)].

Theorem D.1. *Let x_ϵ be a bounded sequence in $L^2(\Omega, \mathbb{R}^m)$. After extracting a subsequence, for which we still keep the index ϵ, there exists a*

Radon measure μ on $T^*\Omega = \Omega \times \mathbb{R}^d$ whose values are nonnegative Hermitian matrices such that the following property holds: Given scalar symbols $p, q \in \mathscr{S}(T^*\Omega, \mathbb{R})$, there is

$$p(t, \epsilon D)x_\epsilon \otimes \overline{q(t, \epsilon D)x_\epsilon} \; \rightharpoonup \; \int_{\mathbb{R}^d} p(t, \tau) \cdot \overline{q(t, \tau)} \, \mu(d\tau) \qquad \text{(D.1)}$$

vaguely in the space of matrix-valued Radon measures. Here, $D = -i\partial$ and the integral denotes the projection of the measure $p\bar{q}\mu$ onto Ω. As a consequence, we obtain the following localization principle: Given a matrix-valued symbol $P \in C^\infty(T^*\Omega, \mathbb{R}^{n \times m})$ that is a polynomial in the dual variable τ, there holds

$$P(t, \epsilon D)x_\epsilon \to 0 \quad \text{strongly in} \quad L^2(\Omega, \mathbb{R}^n) \quad \Longleftrightarrow \quad P\mu = 0. \quad \text{(D.2)}$$

We now turn back to the setting of §II.2, i.e., the proof of the homogenization result Theorem II.1. Let μ be a semiclassical measure belonging to the sequence $\eta_\epsilon = \epsilon^{-1} z_\epsilon$, which is bounded in $L^\infty([0, T], \mathbb{R}^r)$. Thus, by the basic property (D.1) we have, weakly* in $L^\infty([0, T], \mathbb{R}^r)$,[73]

$$\Sigma_\epsilon G_\epsilon^{-1} = \eta_\epsilon \otimes \eta_\epsilon \; \overset{*}{\rightharpoonup} \; \Sigma_0 G_0^{-1} = \int_{\mathbb{R}} \mu(d\tau),$$

$$\Pi_\epsilon G_\epsilon^{-1} = \dot{z}_\epsilon \otimes \dot{z}_\epsilon = \epsilon D_t \eta_\epsilon \otimes \overline{\epsilon D_t \eta_\epsilon} \; \overset{*}{\rightharpoonup} \; \Pi_0 G_0^{-1} = \int_{\mathbb{R}} \tau^2 \, \mu(d\tau).$$

The differential equation (II.25) for the normal motion can be rewritten as

$$(-\epsilon^2 D_t^2 + H(y_0))\eta_\epsilon = O(\epsilon) \quad \text{in} \quad L^\infty([0, T], \mathbb{R}^r).$$

Accordingly, by the localization principle (D.2),

$$(-\tau^2 I + H(y_0))\mu = 0.$$

Integration with respect to the dual variable τ and multiplication by G_0 from the right yields

$$\int_{\mathbb{R}} \tau^2 \, \mu(d\tau) = H(y_0) \int_{\mathbb{R}} \mu(d\tau), \quad \text{resp.} \quad \Pi_0 = H(y_0)\Sigma_0,$$

i.e., the weak virial theorem, Eq. (II.27).

This new proof of the weak virial theorem, though surely far too complicated for the purposes of Chapter II, tells us two interesting things: first, the result is by no means an accident but systematically connected with the differential equation (II.25) of the normal motion, and second, it is but one example of a whole family of similar results. For instance, the corresponding equi-partitioning of energy in Example I.1 is a direct consequence of the div-curl lemma of compensated compactness theory, which can likewise be obtained by the localization principle of H-measures [95, §1.3][37, §2].

---

[73]Strictly speaking, before applying (D.1) we first have to cut-off infinity in the dual variable; afterwards we remove the cut-off by taking a limit: Choose $\chi \in C_c^\infty(\mathbb{R})$ equal to 1 in a neighborhood of 0 and put $\chi_k(\tau) = \chi(\tau/k)$. Now, replace η_ϵ by $\chi_k(\epsilon D_t)\eta_\epsilon$, apply (D.1), and take the limit $k \to \infty$.

Bibliography

[1] R. ABRAHAM AND J. E. MARSDEN, *Foundations of Mechanics*, Addison-Wesley Publ. Co., Redwood City, New York, Bonn, 1985 printing of the 2nd ed., 1985.

[2] H. ALFVÉN, *Cosmical Electrodynamics*, Clarendon Press, Oxford, 1950.

[3] V. I. ARNOLD, *Small denominators and problems of stability of motion in classical and celestial mechanics*, Russ. Math. Surv., 18 (1963), pp. 85–192.

[4] ———, *Lectures on bifurcations in versal families*, Russ. Math. Surv., 27 (1972), pp. 54–123.

[5] ———, *Mathematical Methods of Classical Mechanics*, Springer-Verlag, Berlin, Heidelberg, New York, 1978.

[6] V. I. ARNOLD, V. V. KOZLOV, AND A. I. NEISHTADT, *Mathematical aspects of classical and celestial mechanics*, in Dynamical Systems III, V. I. Arnold, ed., Springer-Verlag, Berlin, Heidelberg, New York, 2nd ed., 1993.

[7] J. E. AVRON, R. SEILER, AND L. G. YAFFE, *Adiabatic theorems and applications to the quantum Hall effect*, Comm. Math. Phys., 110 (1987), pp. 33–49.

[8] ———, *Erratum: Adiabatic theorems and applications to the quantum Hall effect*, Comm. Math. Phys., 156 (1993), pp. 649–650.

[9] P. BALA, P. GROCHOWSKI, B. LESYNG, AND J. A. McCAMMON, *Quantum–classical molecular dynamics. Models and applications*, in Quantum Mechanical Simulation Methods for Studying Biological Systems, M. Fields, ed., Les Houches, France, 1995.

[10] P. BALA, B. LESYNG, AND J. A. McCAMMON, *Extended Hellmann–Feynman theorem for non–stationary states and its application in quantum–classical molecular dynamics simulation*, Chem. Phys. Lett., 219 (1994), pp. 259–266.

[11] G. BENETTIN, L. GALGANI, AND A. GIORGILLI, *Realization of holonomic constraints and freezing of high frequency degrees of freedom in the light of classical perturbation theory. I*, Comm. Math. Phys., 113 (1987), pp. 87–103.

[12] ———, *Realization of holonomic constraints and freezing of high frequency degrees of freedom in the light of classical perturbation theory. II*, Comm. Math. Phys., 121 (1989), pp. 557–601.

[13] A. BENSOUSSAN, J.-L. LIONS, AND G. PAPANICOLAOU, *Asymptotic Analysis for Periodic Structures*, North-Holland Publ. Co., Amsterdam, New York, 1978.

[14] H. J. C. BERENDSEN AND J. MAVRI, *Quantum simulation of reaction dynamics by density matrix evolutions*, J. Phys. Chem., 97 (1993), pp. 13464–13468.

[15] J. BERKOWITZ AND C. S. GARDNER, *On the asymptotic series expansion of the motion of a charged particle in slowly varying fields*, Comm. Pure Appl. Math., 12 (1959), pp. 501–512.

[16] M. BORN AND V. FOCK, *Beweis des Adiabatensatzes*, Z. Phys., 51 (1928), pp. 165–180.

[17] F. A. BORNEMANN, P. NETTESHEIM, AND C. SCHÜTTE, *Quantum-classical molecular dynamics as an approximation to full quantum dynamics*, J. Chem. Phys., 105 (1996), pp. 1074–1083.

[18] F. A. BORNEMANN AND C. SCHÜTTE, *Homogenization of Hamiltonian systems with a strong constraining potential*, Physica D, 102 (1997), pp. 57–77.

[19] ——, *On the singular limit of the quantum-classical molecular dynamics model*, Preprint SC-97-7, Konrad-Zuse-Zentrum, Berlin, 1997. (submitted to SIAM J. Appl. Math.).

[20] R. BOTT, *Nondegenerate critical manifolds*, Ann. Math., 60 (1954), pp. 248–261.

[21] J. Y. CHEMIN, *A propos d'un problème de pénalisation de type anti symétrique*, C.R.A.S. Paris Ser. I, 321 (1995), pp. 861–864.

[22] J. M. COMBES, *The Born-Oppenheimer approximation*, Acta Phys. Austriaca, 17, Suppl., (1977), pp. 139–159.

[23] B. DACOROGNA, *Weak Continuity and Weak Lower Semicontinuity of Non-Linear Functionals*, Springer-Verlag, Berlin, Heidelberg, New York, 1982.

[24] J. DIESTEL AND J. J. UHL, *Vector Measures*, Amer. Math. Soc., Providence, Rhode Island, 1977.

[25] N. DUNFORD AND J. T. SCHWARTZ, *Linear Operators Part I: General Theory*, John Wiley, New York, Chichester, 1957.

[26] D. G. EBIN, *The motion of slightly compressible fluids viewed as a motion with strong constraining force*, Ann. Math., 105 (1977), pp. 141–200.

[27] P. F. EMBID AND A. J. MAJDA, *Averaging over fast gravity waves for geophysical flows with arbitrary potential vorticity*, Comm. Part. Diff. Eq., 21 (1996), pp. 619–658.

[28] L. C. EVANS, *Weak Convergence Methods for Nonlinear Partial Differential Equations*, vol. 74 of CBMS, Regional Conference Series in Mathematics, Amer. Math. Soc., Providence, Rhode Island, 1990, 2nd printing 1993.

[29] L. C. EVANS AND R. F. GARIEPY, *Measure Theory and Fine Properties of Functions*, CRC Press, Inc., Boca Raton, Ann Arbor, London, 1992.

[30] U. FANO, *Description of states in quantum mechanics by density matrix and operator techniques*, Rev. Mod. Phys., 29 (1957), pp. 74–93.

[31] F. M. FERNÁNDEZ AND E. A. CASTRO, *Hypervirial Theorems*, Lecture Notes in Chemistry, Springer-Verlag, Berlin, Heidelberg, New York, 1987.

[32] K.-O. FRIEDRICHS, *On the adiabatic theorem in quantum theory. Part I*, Report IMM-NYU-218, New York University, 1955.

[33] ——, *On the adiabatic theorem in quantum theory. Part II*, Report IMM-NYU-230, New York University, 1955.

[34] G. GALLAVOTTI, *The Elements of Mechanics*, Springer-Verlag, Berlin, Heidelberg, New York, 1983.

[35] A. GARCIA-VELA, R. GERBER, AND D. IMRE, *Mixed quantum wave packet/classical trajectory treatment of the photodissociation process ArHCl to Ar+H+Cl*, J. Chem. Phys., 97 (1992), pp. 7242–7250.

[36] P. GÉRARD, *Mesures semi-classiques et ondes de Bloch*, Semin. Equ. Deriv. Partielles, Ecole Polytechnique, Exposé XVI (1991). 18 pp.

[37] ——, *Microlocal defect measures*, Comm. Part. Diff. Eq., 16 (1991), pp. 1761–1794.

[38] D. GILBARG AND N. S. TRUDINGER, *Elliptic Partial Differential Equations of Second Order*, Springer-Verlag, Berlin, Heidelberg, New York, 2nd ed., 1983.

[39] H. GOLDSTEIN, *Classical Mechanics*, Addison-Wesley Publ. Co., Cambridge, 1953.

[40] G. A. HAGEDORN, *Semiclassical quantum mechanics I*, Comm. Math. Phys., 71 (1980), pp. 77–93.

[41] ———, *A time dependent Born-Oppenheimer approximation*, Comm. Math. Phys., 77 (1980), pp. 1–19.

[42] ———, *Electron energy level crossing in the time–dependent Born-Oppenheimer approximation*, Theor. Chim. Acta, 77 (1990), pp. 163–190.

[43] ———, *Molecular propagation through electron energy level crossings*, Mem. Amer. Math. Soc., 536 (1994), pp. 1–130.

[44] G. HELLWIG, *Über die Bewegung geladener Teilchen in schwach veränderlichen Magnetfeldern*, Z. Naturforschung, 10 (1955), pp. 508–516.

[45] E. HEWITT AND K. R. STROMBERG, *Real and Abstract Analysis; A Modern Treatment of the Theory of Functions of a Real Variable*, Springer-Verlag, Berlin, New York, 1965.

[46] J. O. HIRSCHFELDER, *Classical and quantum mechanical hypervirial theorems*, J. Chem. Phys., 33 (1960), pp. 1462–1466.

[47] V. V. JIKOV, S. M. KOZLOV, AND O. A. OLEINIK, *Homogenization of Differential Operators and Integral Functionals*, Springer-Verlag, Berlin, Heidelberg, New York, 1994.

[48] E. KAMKE, *Zur Theorie der Systeme gewöhnlicher Differentialgleichungen, II*, Acta Math., 58 (1932), pp. 57–85.

[49] N. G. V. KAMPEN, *Elimination of fast variables*, Phys. Rep., 124 (1985), pp. 69–160.

[50] T. KATO, *On the adiabatic theorem of quantum mechanics*, J. Phys. Soc. Jap., 5 (1950), pp. 435–439.

[51] ———, *Perturbation Theory for Linear Operators*, Springer-Verlag, Berlin, Heidelberg, New York, 2nd ed., 1984.

[52] J. B. KELLER AND J. RUBINSTEIN, *Nonlinear wave motion in a strong potential*, Wave Motion, 13 (1991), pp. 291–302.

[53] J. KEVORKIAN AND J. D. COLE, *Multiple Scale and Singular Perturbation Methods*, Springer-Verlag, Berlin, Heidelberg, New York, 1996.

[54] J. KISYŃSKI, *Sur les opérateurs de Green des problèmes de Cauchy abstraits*, Studia Math., 23 (1963), pp. 285–328.

[55] S. KLAINERMAN AND A. J. MAJDA, *Singular limits of quasilinear hyperbolic systems with large parameters and the incompressible limit of compressible fluids*, Comm. Pure Appl. Math., 43 (1981), pp. 481–524.

[56] H. KNESER, *Über die Lösungen eines Systems gewöhnlicher Differentialgleichungen, das der Lipschitzschen Bedingung nicht genügt*, Sitz.-Ber. Preuss. Akad. Wiss. Phys.-Math. Kl., (1923), pp. 171–174.

[57] J. KOILLER, *A note on classical motions under strong constraints*, J. Phys. A: Math. Gen., 23 (1990), pp. L521–L527.

[58] H. KOPPE AND H. JENSEN, *Das Prinzip von d'Alembert in der Klassischen Mechanik und in der Quantenmechanik*, Sitz.-Ber. Heidelb. Akad. Wiss. Math.-Naturwiss. Kl., 5. Abh., (1971).

[59] H.-O. KREISS, *Problems with different time scales*, in Acta Numerica 1992, A. Iserles, ed., Cambridge University Press, Cambridge, 1992.

[60] M. KRUSKAL, *The gyration of a charged particle*, Rept. PM-S-33 (NYO-7903), Princeton University, Project Matterhorn, 1958.

[61] L. D. LANDAU AND E. M. LIFSHITZ, *The Classical Theory of Fields*, Pergamon Press, Oxford, London, New York, Paris, 2nd ed., 1962.

[62] ——, *Mechanics*, Pergamon Press, Oxford, London, New York, Paris, 1965.

[63] E. H. LIEB AND M. LOSS, *Analysis*, Amer. Math. Soc., Providence, Rhode Island, 1997.

[64] J.-L. LIONS, *Équations différentielles opérationnelles et problèmes aux limites*, Springer-Verlag, Berlin, Heidelberg, New York, 1961.

[65] P.-L. LIONS AND T. PAUL, *Sur les mesures de Wigner*, Rev. Mat. Iberoamer., 9 (1993), pp. 261–270.

[66] C. LUBICH, *Integration of stiff mechanical systems by Runge-Kutta methods*, Z. angew. Math. Phys., 44 (1993), pp. 1022–1053.

[67] A. J. MAJDA, *Nonlinear geometric optics for hyperbolic systems of conservation laws*, in Oscillation Theory, Computation, and Methods of Compensated Compactness, C. Dafermos, J. Erikson, D. Kinderlehrer, and I. Müller, eds., vol. 2 of IMA, Springer-Verlag, New York, 1986, pp. 115–165.

[68] V. P. MASLOV AND M. V. FEDORIUK, *Semi-Classical Approximation in Quantum Mechanics*, D. Reidel Publishing Company, Dordrecht, Boston, London, 1981.

[69] A. MESSIAH, *Quantum Mechanics. Vol. I & II*, North-Holland Publ. Co., Amsterdam, New York, 1962.

[70] F. MURAT AND L. TARTAR, *H-convergence*, in Topics in the Mathematical Modeling of Composite Materials, R. V. Kohn, ed., Birkhäuser, Boston, Basel, 1994.

[71] J. V. NEUMANN AND E. WIGNER, *Über das Verhalten von Eigenwerten bei adiabatischen Prozessen*, Phys. Zeit., 30 (1929), pp. 467–470.

[72] T. G. NORTHROP, *The Adiabatic Motion of Charged Particles*, Interscience Publishers, New York, London, Sydney, 1963.

[73] A. PAZY, *Semigroups of Linear Operators and Applications to Partial Differential Equations*, Springer-Verlag, Berlin, Heidelberg, New York, 1983.

[74] C. PUGH, *Funnel sections*, J. Diff. Eqs., 19 (1975), pp. 270–295.

[75] J. RAUCH, *Lectures on geometric optics*. Notes of a Minicourse at the IAS-Park City Mathematics Institute, 1995. URL = http://www.math.lsa.umich.edu/~rauch.

[76] M. REED AND B. SIMON, *Methods of Modern Mathematical Physics II. Fourier Analysis, Self-Adjointness*, Academic Press, New York, 1975.

[77] ——, *Methods of Modern Mathematical Physics IV. Analysis of Operators*, Academic Press, New York, 1978.

[78] ——, *Methods of Modern Mathematical Physics I. Functional Analysis*, Academic Press, New York, 2nd, rev. ed., 1980.

[79] S. REICH, *Smoothed dynamics of highly oscillatory Hamiltonian systems*, Physica D, 89 (1995), pp. 28–42.

[80] F. RELLICH, *Störungstheorie der Spektralzerlegung, I*, Math. Ann., 113 (1937), pp. 600–619.

[81] J. R. RINGROSE, *Compact non-self-adjoint operators*, Van Nostrand Reinhold Co., London, New York, 1971.

[82] H. RUBIN AND P. UNGAR, *Motion under a strong constraining force*, Comm. Pure Appl. Math., 10 (1957), pp. 65–87.

[83] W. RUDIN, *Functional Analysis*, McGraw-Hill Publishing Co., New York, London, 1973.

[84] ——, *Real and Complex Analysis*, McGraw-Hill Publishing Co., New York, London, Sydney, Toronto, 1987.

[85] R. SCHATTEN, *Norm ideals of completely continuous operators*, Springer-Verlag, Berlin, New York, 1970.

[86] H.-J. SCHMIDT, *Models for constrained motion and d'Alembert's principle*, Z. angew. Math. Mech., 73 (1993), pp. 155–163.

[87] S. SCHOCHET, *Fast singular limits of hyperbolic partial differential equations*, J. Diff. Eq., 114 (1994), pp. 476–512.

[88] C. SCHÜTTE AND F. A. BORNEMANN, *Approximation properties and limits of the quantum-classical molecular dynamics model*, in Algorithms for Macromolecular Modelling, P. Deuflhard, J. Hermans, B. Leimkuhler, A. Mark, S. Reich, and R. D. Skeel, eds., Springer-Verlag, Berlin, Heidelberg, New York, to appear.

[89] ———, *Homogenization approach to smoothed molecular dynamics*, Nonlinear Analysis, (to appear).

[90] I. SEGAL, *Non-linear semi-groups*, Ann. Math., 78 (1963), pp. 339–364.

[91] B. SIMON, *Quantum Mechanics for Hamiltonians Defined as Quadratic Forms*, Princeton University Press, Princeton, NJ, 1971.

[92] ———, *Trace ideals and their applications*, Cambridge University Press, Cambridge, London, New York, 1979.

[93] L. SPITZER, *Physics of Fully Ionized Gases*, Interscience Publishers, New York, London, Sydney, 1956.

[94] F. TAKENS, *Motion under the influence of a strong constraining force*, in Global Theory of Dynamical Systems, Evanston 1979, Z. Nitecki and C. Robinson, eds., Springer-Verlag, Berlin, Heidelberg, New York, 1980, pp. 425–445.

[95] L. TARTAR, *H-measures, a new approach for studying homogenisation, oscillations and concentration effects in partial differential equations*, Proc. R. Soc. Edinb., Sect. A, 115 (1990), pp. 193–230.

[96] ———, *H-measures and applications*, in Proceedings of the international congress of mathematicians (ICM), August 21-29, 1990, Kyoto, Japan, vol. II, Springer-Verlag, Berlin, Heidelberg, New York, 1991, pp. 1215–1223.

[97] ———, *Beyond Young measures*, Meccanica, 30 (1995), pp. 505–526.

[98] A. C. ZAANEN, *An Introduction to the Theory of Integration*, North-Holland Publ. Co., Amsterdam, 1958.

[99] G. ZHANG AND T. SCHLICK, *LIN: A new algorithm to simulate the dynamics of biomolecules by combining implicit integration and normal mode technique*, J. Comp. Chem., 14 (1993), pp. 1212–1233.

List of Symbols

Function Spaces

Ω	domain in $\mathbb{R}^d$
$\mathscr{X}$	Banach space
$C^k(\overline{\Omega}, \mathscr{X})$	k-times continuously differentiable $f : \Omega \to \mathscr{X}$, all derivatives have a bounded maximum norm
$C^{k,1}(\overline{\Omega}, \mathscr{X})$	$C^k(\overline{\Omega}, \mathscr{X})$, kth derivative is uniformly Lipschitz
$L^p(\Omega, \mathscr{X})$	Lebesgue-Bochner spaces, p. 119
$L^p(\Omega)$	$L^p(\Omega, \mathbb{R})$ or $L^p(\Omega, \mathbb{C})$
$H^1(\Omega), H^{-1}(\Omega), W^{1,\infty}(\Omega)$	Sobolev spaces
$\mathscr{R}$	Rollnik class, p. 129

Operator Spaces

$\mathscr{H}$	separable Hilbert space
$\langle \cdot, \cdot \rangle$	inner product on $\mathscr{H}$
$\mathscr{H}_+ \hookrightarrow \mathscr{H} \hookrightarrow \mathscr{H}_-$	rigging of $\mathscr{H}$, p. 124
$\sigma : \mathscr{H}_- \to \mathscr{H}_+$	Riesz representation map of the rigging, p. 125
$A^\sigma, {}^\sigma A$	p. 125
$\mathscr{B}(\mathscr{H})$	bounded linear operators on $\mathscr{H}$
$\mathscr{B}(\mathscr{H}_1, \mathscr{H}_2)$	bounded linear operators from $\mathscr{H}_1$ to $\mathscr{H}_2$
$\mathscr{K}(\mathscr{H})$	compact linear operators on $\mathscr{H}$
$\mathscr{J}_1(\mathscr{H})$	trace class operators on $\mathscr{H}$, p. 117
$\mathrm{tr}\, A$	trace of $A \in \mathscr{J}_1(\mathscr{H})$, p. 118

Differential Geometry

TM, TN	tangent bundle of M, resp. $N \subset M$
$TN^\perp$	normal bundle of $N \subset M$
$TM|N$	TM restricted to base points in $N \subset M$
$\langle \cdot, \cdot \rangle$	Riemannian metric, p. 17
G	metric tensor, p. 29
∇	Levi-Cività connection, p. 18
$\mathrm{grad}\, U$	gradient of U with respect $\langle \cdot, \cdot \rangle$, p. 29
Γ, Γ^i_{jk}	Christoffel symbols, p. 29
$\hat{\Gamma}$	nonsymmetric variant of Christoffel symbols, p. 29
$\hat{\Gamma} : A$	$(\hat{\Gamma} : A)^i = \hat{\Gamma}^i_{jk} A^{jk}$, p. 35
$\exp_y X$	geodesic exponential function, p. 31

Homogenization of Mechanical Systems

$\mathscr{L}_\epsilon$	Lagrangian, singularly perturbed system, p. 18
$\mathscr{L}_{\text{hom}}$	Lagrangian, homogenized system, p. 22
$\mathscr{L}_{\text{con}}$	Lagrangian, "realization-of-constraints" system, p. 51
$\mathscr{L}_{\text{hom}}^\kappa$	Lagrangian, homogenized system with friction, p. 68
U_{hom}	homogenization of the potential U, p. 21
U_{BO}	Born-Oppenheimer potential, p. 106
H, H_U	Hessian of constraining potential U, p. 19
	Hamiltonian, pp. 91 and 106
ω, ω_λ	normal frequencies of Hessian H, p. 20,
	energy levels of Hamiltonian H, pp. 91, 99, and 106
$\theta_\epsilon, \theta_\epsilon^\lambda$	adiabatic "invariants" (actions, energy level probabilities)
$\theta_0, \theta_0^\lambda$	limits of the adiabatic "invariants"
$\theta_*, \theta_*^\lambda$	initial values of the limits $\theta_0, \theta_0^\lambda$
F_U, F_V	force term corresponding to U, resp. V, p. 29
x_*, v_*	initial position and velocity, pp. 19 and 51
y, z	coordinate splitting $x = y + z$, $x = \exp_y z$, p. 31
P, P_λ, Q	projections, pp. 20, 91, 99, 106
$\Sigma_\epsilon, \Pi_\epsilon$	quadratic terms of normal components, p. 34
f_ϵ, H_ϵ, etc.	$f_\epsilon = f(y_\epsilon)$, $H_\epsilon = H(y_\epsilon)$, etc., pp. 32 and 33
σ_λ	$\sigma_\lambda = \text{tr}(P_{0\lambda} \Sigma_0 P_{0\lambda})$, p. 45
ρ_ϵ	time-dependent density operator, pp. 95 and 109
$T_\epsilon^\parallel, U_\epsilon^\parallel, E_\epsilon^\parallel$	energies, constrained motion, p. 36
$T_\epsilon^\perp, U_\epsilon^\perp, E_\epsilon^\perp$	energies, normal motion, p. 37
$T_{\epsilon\lambda}^\perp, U_{\epsilon\lambda}^\perp, E_{\epsilon\lambda}^\perp$	energies, components of normal motion, p. 44
$\text{grad}\, H$	$(\text{grad}\, H)_k^{ij} = g^{il} \dfrac{\partial H_l^j}{\partial x^k}$, p. 35
$\text{grad}\, H : B$	$(\text{grad}\, H : B)^i = (\text{grad}\, H)_k^{ij} B_j^k$, p. 35

Index

General Remarks

Lecture Notes are printed by photo-offset from the master-copy delivered in camera-ready form by the authors. For this purpose Springer-Verlag provides technical instructions for the preparation of manuscripts.

Careful preparation of manuscripts will help keep production time short and ensure a satisfactory appearance of the finished book. The actual production of a Lecture Notes volume normally takes approximately 8 weeks.

Authors receive 50 free copies of their book. No royalty is paid on Lecture Notes volumes.

Authors are entitled to purchase further copies of their book and other Springer mathematics books for their personal use, at a discount of 33,3 % directly from Springer-Verlag.

Commitment to publish is made by letter of intent rather than by signing a formal contract. Springer-Verlag secures the copyright for each volume.

Addresses:

Professor A. Dold
Mathematisches Institut
Universität Heidelberg
Im Neuenheimer Feld 288
D-69120 Heidelberg, Germany

Professor F. Takens
Mathematisch Instituut
Rijksuniversiteit Groningen
Postbus 800
NL-9700 AV Groningen
The Netherlands

Professor Bernard Teissier
École Normale Supérieure
45, rue d'Ulm
F-7500 Paris, France

Springer-Verlag, Mathematics Editorial
Tiergartenstr. 17
D-69121 Heidelberg, Germany
Tel.: *49 (6221) 487-410